I LOVE MY FOOD AND EXERCISE JOURNAL

Food Journals Publishing

I Love my Food and Exercise Journal

by Food Journals Publishing

Food Journals Publishing

ISBN-13: 978-1499735505
ISBN-10: 1499735502

As well as changing **what** you eat, its important to eat the correct **quantities**; aim for small, healthy meals. Also, when you look up how many calories a particular food contains, the figure will depend on the portion size.

A lot of the foods that you eat will state the number of calories on the packaging. There are also lots of online resources and books available which give the calorie values of common foods.

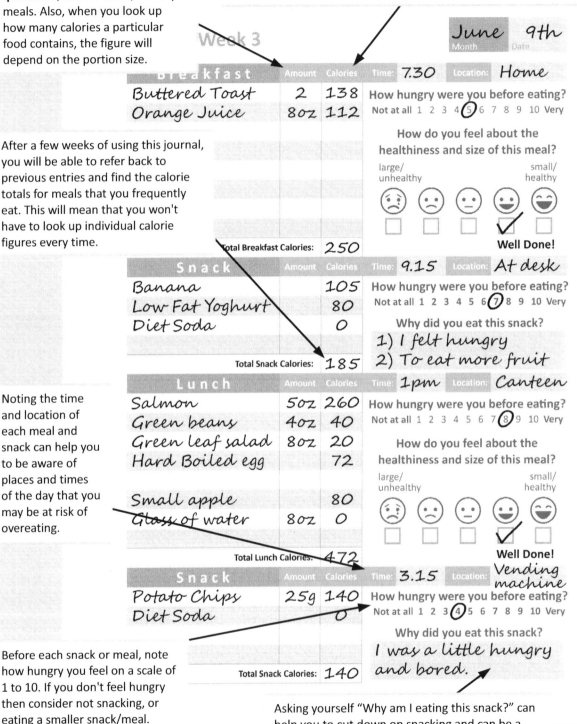

Week 3

June 9th
Month   Date

Breakfast
Time: 7.30   Location: Home

	Amount	Calories
Buttered Toast	2	138
Orange Juice	8oz	112

How hungry were you before eating?
Not at all 1 2 3 4 ⑤ 6 7 8 9 10 Very

How do you feel about the healthiness and size of this meal?

large/unhealthy   small/healthy

Total Breakfast Calories: 250

Well Done!

After a few weeks of using this journal, you will be able to refer back to previous entries and find the calorie totals for meals that you frequently eat. This will mean that you won't have to look up individual calorie figures every time.

Snack
Time: 9.15   Location: At desk

	Amount	Calories
Banana		105
Low Fat Yoghurt		80
Diet Soda		0

Total Snack Calories: 185

How hungry were you before eating?
Not at all 1 2 3 4 5 6 ⑦ 8 9 10 Very

Why did you eat this snack?
1) I felt hungry
2) To eat more fruit

Lunch
Time: 1pm   Location: Canteen

	Amount	Calories
Salmon	5oz	260
Green beans	4oz	40
Green leaf salad	8oz	20
Hard Boiled egg		72
Small apple		80
~~Glass~~ of water	8oz	0

How hungry were you before eating?
Not at all 1 2 3 4 5 6 7 ⑧ 9 10 Very

How do you feel about the healthiness and size of this meal?

large/unhealthy   small/healthy

Total Lunch Calories: 472

Well Done!

Noting the time and location of each meal and snack can help you to be aware of places and times of the day that you may be at risk of overeating.

Snack
Time: 3.15   Location: Vending machine

	Amount	Calories
Potato Chips	25g	140
Diet Soda		0

Total Snack Calories: 140

How hungry were you before eating?
Not at all 1 2 3 ④ 5 6 7 8 9 10 Very

Why did you eat this snack?
I was a little hungry and bored.

Before each snack or meal, note how hungry you feel on a scale of 1 to 10. If you don't feel hungry then consider not snacking, or eating a smaller snack/meal.

Asking yourself "Why am I eating this snack?" can help you to cut down on snacking and can be a powerful tool in helping you to identify possible emotional reasons for your eating behavior.

We are all advised to drink eight 8oz glasses of fluid each day, and at least 5 portions of fruit and vegetables. Use these handy charts to tick off your fluid and fruit and vegetable intake each day.

You probably won't fill this section in every day, however you should weigh at least once a week in order to monitor your progress. Each time you weigh, try to do so at the same time (such as just before Dinner) as your weight can fluctuate over the course of a day.

Weight	178 lb
Time	6.20 pm

Dinner

	Amount	Calories	Time: 6.30	Location: Home
Chicken Salad	9oz	125		
Pita bread	1	160		
Glass of water	8oz	0		

How hungry were you before eating?
Not at all 1 2 3 4 5 6 ⑦ 8 9 10 Very

How do you feel about the healthiness and size of this meal?

large/unhealthy   small/healthy

(selected: small/healthy) ✓ **Well Done!**

Total Dinner Calories: 285

Snack

	Amount	Calories	Time: 9.30	Location: Kitchen
Bowl of cereal	8oz	100		
1% Milk	8oz	105		

How hungry were you before eating?
Not at all 1 2 3 4 5 6 ⑦ 8 9 10 Very

Why did you eat this snack?

I felt hungry

Total Snack Calories: 205

Total Daily Calories: 1537

Exercise	Target	Achievement
Swim with Jayne	15 lengths	12 lengths

The 'smiley faces' feature of this journal is designed to help you to develop a better mindset and attitude towards food: small/healthy meals make you happy, whereas large/unhealthy meals make you sad.

As well as helping you to lose weight, regular exercise helps to combat health problems, relieve stress and boost your energy levels. Find an activity that you enjoy, and aim for at least 30 minutes of physical activity each day.

Are you happy with how you ate and exercised today?

Food	Exercise
(4th face selected) ✓ Well Done!	(4th face selected) ✓ Well Done!

Your Starting Statistics

○ ○ ○ ○ ○ ○ ○ ○ ○ ○ ○

Enter your starting statistics into the table below, along with your realistic aims. You don't have to fill in everything, plus, there's room for you to add other measurements if you wish.

Once you've completed this 10-week Journal, return to this page and enter your final statistics into the table. This will help you to see how close you got to your aims.

Date:

	At the Start (Your current statistics)	Your Aims (Set achievable goals)	After 10 weeks (Fill this in at the end)
Weight:			
Chest:			
Waist:			
Hips:			
Thighs:			
Calves:			
Upper arms:			
Cholesterol:			
Blood pressure:			

Week 1

Monday Week 1

Breakfast	Amount	Calories	Time:	Location:

How hungry were you before eating?

Not at all 1 2 3 4 5 6 7 8 9 10 Very

How do you feel about the healthiness and size of this meal?

large/ unhealthy   small/ healthy

☐ ☐ ☐ ☐ ☐

Well Done!

Total Breakfast Calories:

Snack	Amount	Calories	Time:	Location:

How hungry were you before eating?

Not at all 1 2 3 4 5 6 7 8 9 10 Very

Why did you eat this snack?

Total Snack Calories:

Lunch	Amount	Calories	Time:	Location:

How hungry were you before eating?

Not at all 1 2 3 4 5 6 7 8 9 10 Very

How do you feel about the healthiness and size of this meal?

large/ unhealthy   small/ healthy

☐ ☐ ☐ ☐ ☐

Well Done!

Total Lunch Calories:

Snack	Amount	Calories	Time:	Location:

How hungry were you before eating?

Not at all 1 2 3 4 5 6 7 8 9 10 Very

Why did you eat this snack?

Total Snack Calories:

| | Weight |
| | Time |

Dinner

	Amount	Calories	Time:	Location:
Total Dinner Calories:				

How hungry were you before eating?
Not at all 1 2 3 4 5 6 7 8 9 10 Very

How do you feel about the healthiness and size of this meal?

large/unhealthy .... small/healthy

☐ ☐ ☐ ☐ ☐

Well Done!

Snack

	Amount	Calories	Time:	Location:
Total Snack Calories:				

How hungry were you before eating?
Not at all 1 2 3 4 5 6 7 8 9 10 Very

Why did you eat this snack?

Total Daily Calories:

Exercise	Target	Achievement

Are you happy with how you ate and exercised today?

Food	Exercise
☐ ☐ ☐ ☐ ☐	☐ ☐ ☐ ☐ ☐
Well Done!	**Well Done!**

Thursday Week 1

Breakfast

	Amount	Calories	Time:	Location:
Total Breakfast Calories:				

How hungry were you before eating?

Not at all 1 2 3 4 5 6 7 8 9 10 Very

How do you feel about the healthiness and size of this meal?

large/unhealthy   small/healthy

☹ ☹ 😐 🙂 😄
☐ ☐ ☐ ☐ ☐

Well Done!

Snack

	Amount	Calories	Time:	Location:
Total Snack Calories:				

How hungry were you before eating?

Not at all 1 2 3 4 5 6 7 8 9 10 Very

Why did you eat this snack?

Lunch

	Amount	Calories	Time:	Location:
Total Lunch Calories:				

How hungry were you before eating?

Not at all 1 2 3 4 5 6 7 8 9 10 Very

How do you feel about the healthiness and size of this meal?

large/unhealthy   small/healthy

☹ ☹ 😐 🙂 😄
☐ ☐ ☐ ☐ ☐

Well Done!

Snack

	Amount	Calories	Time:	Location:
Total Snack Calories:				

How hungry were you before eating?

Not at all 1 2 3 4 5 6 7 8 9 10 Very

Why did you eat this snack?

	Weight	
	Time	

D i n n e r

	Amount	Calories	Time:	Location:
Total Dinner Calories:				

How hungry were you before eating?
Not at all 1 2 3 4 5 6 7 8 9 10 Very

How do you feel about the healthiness and size of this meal?

large/unhealthy   small/healthy

☹ 😕 😐 😃 😄

☐ ☐ ☐ ☐ ☐

Well Done!

S n a c k

	Amount	Calories	Time:	Location:
Total Snack Calories:				

How hungry were you before eating?
Not at all 1 2 3 4 5 6 7 8 9 10 Very

Why did you eat this snack?

Total Daily Calories:

E x e r c i s e	Target	Achievement

Are you happy with how you ate and exercised today?

F o o d	E x e r c i s e
☹ 😕 😐 😃 😄	☹ 😕 😐 😃 😄
☐ ☐ ☐ ☐ ☐	☐ ☐ ☐ ☐ ☐
Well Done!	**Well Done!**

Friday Week 1

Breakfast

	Amount	Calories	Time:	Location:

Total Breakfast Calories:

How hungry were you before eating?

Not at all 1 2 3 4 5 6 7 8 9 10 Very

How do you feel about the healthiness and size of this meal?

large/unhealthy   small/healthy

☹ 🙁 😐 😃 😄
☐ ☐ ☐ ☐ ☐

Well Done!

Snack

	Amount	Calories	Time:	Location:

Total Snack Calories:

How hungry were you before eating?

Not at all 1 2 3 4 5 6 7 8 9 10 Very

Why did you eat this snack?

Lunch

	Amount	Calories	Time:	Location:

Total Lunch Calories:

How hungry were you before eating?

Not at all 1 2 3 4 5 6 7 8 9 10 Very

How do you feel about the healthiness and size of this meal?

large/unhealthy   small/healthy

☹ 🙁 😐 😃 😄
☐ ☐ ☐ ☐ ☐

Well Done!

Snack

	Amount	Calories	Time:	Location:

Total Snack Calories:

How hungry were you before eating?

Not at all 1 2 3 4 5 6 7 8 9 10 Very

Why did you eat this snack?

	Weight	
	Time	

Dinner

	Amount	Calories	Time:	Location:
Total Dinner Calories:				

How hungry were you before eating?
Not at all 1 2 3 4 5 6 7 8 9 10 **Very**

How do you feel about the healthiness and size of this meal?

large/
unhealthy   small/
healthy

☹ ☹ 😐 😀 😄
☐ ☐ ☐ ☐ ☐

Well Done!

Snack

	Amount	Calories	Time:	Location:
Total Snack Calories:				

How hungry were you before eating?
Not at all 1 2 3 4 5 6 7 8 9 10 **Very**

Why did you eat this snack?

Total Daily Calories:

Exercise	Target	Achievement

Are you happy with how you ate and exercised today?

Food	Exercise
☹ ☹ 😐 😀 😄	☹ ☹ 😐 😀 😄
☐ ☐ ☐ ☐ ☐	☐ ☐ ☐ ☐ ☐
Well Done!	**Well Done!**

Saturday Week 1

Breakfast	Amount	Calories	Time:	Location:

How hungry were you before eating?

Not at all 1 2 3 4 5 6 7 8 9 10 Very

How do you feel about the healthiness and size of this meal?

large/ unhealthy   small/ healthy

☹ ☹ 😐 😀 😄
▢ ▢ ▢ ▢ ▢

Well Done!

Total Breakfast Calories:

Snack	Amount	Calories	Time:	Location:

How hungry were you before eating?

Not at all 1 2 3 4 5 6 7 8 9 10 Very

Why did you eat this snack?

Total Snack Calories:

Lunch	Amount	Calories	Time:	Location:

How hungry were you before eating?

Not at all 1 2 3 4 5 6 7 8 9 10 Very

How do you feel about the healthiness and size of this meal?

large/ unhealthy   small/ healthy

☹ ☹ 😐 😀 😄
▢ ▢ ▢ ▢ ▢

Well Done!

Total Lunch Calories:

Snack	Amount	Calories	Time:	Location:

How hungry were you before eating?

Not at all 1 2 3 4 5 6 7 8 9 10 Very

Why did you eat this snack?

Total Snack Calories:

				Weight	
				Time	

D i n n e r

	Amount	Calories	Time:	Location:

How hungry were you before eating?

Not at all 1 2 3 4 5 6 7 8 9 10 Very

How do you feel about the healthiness and size of this meal?

large/unhealthy   small/healthy

☹ ☹ 😐 🙂 😄

☐ ☐ ☐ ☐ ☐

Well Done!

Total Dinner Calories:

S n a c k

	Amount	Calories	Time:	Location:

How hungry were you before eating?

Not at all 1 2 3 4 5 6 7 8 9 10 Very

Why did you eat this snack?

Total Snack Calories:

Total Daily Calories:

E x e r c i s e	Target	Achievement

Are you happy with how you ate and exercised today?

F o o d	E x e r c i s e

☹ ☹ 😐 🙂 😄   ☹ ☹ 😐 🙂 😄

☐ ☐ ☐ ☐ ☐   ☐ ☐ ☐ ☐ ☐

Well Done!   **Well Done!**

Sunday Week 1

Breakfast	Amount	Calories	Time:	Location:
Total Breakfast Calories:				

How hungry were you before eating?
Not at all 1 2 3 4 5 6 7 8 9 10 Very

How do you feel about the healthiness and size of this meal?

large/
unhealthy   small/
                                             healthy

☹ 🙁 😐 🙂 😄
☐ ☐ ☐ ☐ ☐

Well Done!

Snack	Amount	Calories	Time:	Location:
Total Snack Calories:				

How hungry were you before eating?
Not at all 1 2 3 4 5 6 7 8 9 10 Very

Why did you eat this snack?

Lunch	Amount	Calories	Time:	Location:
Total Lunch Calories:				

How hungry were you before eating?
Not at all 1 2 3 4 5 6 7 8 9 10 Very

How do you feel about the healthiness and size of this meal?

large/
unhealthy   small/
                                             healthy

☹ 🙁 😐 🙂 😄
☐ ☐ ☐ ☐ ☐

Well Done!

Snack	Amount	Calories	Time:	Location:
Total Snack Calories:				

How hungry were you before eating?
Not at all 1 2 3 4 5 6 7 8 9 10 Very

Why did you eat this snack?

| | | | | | Weight | |
| | | | | | Time | |

Dinner

	Amount	Calories	Time:	Location:

Total Dinner Calories:

How hungry were you before eating?
Not at all 1 2 3 4 5 6 7 8 9 10 **Very**

How do you feel about the healthiness and size of this meal?

large/ unhealthy · · · small/ healthy

☹ 😐 😐 😀 😄
☐ ☐ ☐ ☐ ☐

Well Done!

Snack

	Amount	Calories	Time:	Location:

Total Snack Calories:

How hungry were you before eating?
Not at all 1 2 3 4 5 6 7 8 9 10 **Very**

Why did you eat this snack?

Total Daily Calories:

Exercise	Target	Achievement

Are you happy with how you ate and exercised today?

Food	Exercise
☹ 😐 😐 😀 😄	☹ 😐 😐 😀 😄
☐ ☐ ☐ ☐ ☐	☐ ☐ ☐ ☐ ☐
Well Done!	**Well Done!**

Your Weekly Progress

◯ ◯ ◯ ◯ ◯ ◯ ◯ ◯ ◯ ◯

Date: _____

	This week's measurements
Weight:	
Chest:	
Waist:	
Hips:	
Thighs:	
Calves:	
Upper arms:	
Cholesterol:	
Blood pressure:	

How do you feel about this week's progress?

Well Done!

Things you did well this week:

Things you can improve:

Week 2

Monday Week 2

Breakfast

	Amount	Calories	Time:	Location:
Total Breakfast Calories:				

How hungry were you before eating?
Not at all 1 2 3 4 5 6 7 8 9 10 Very

How do you feel about the healthiness and size of this meal?

large/ unhealthy   small/ healthy

☐ ☐ ☐ ☐ ☐

Well Done!

Snack

	Amount	Calories	Time:	Location:
Total Snack Calories:				

How hungry were you before eating?
Not at all 1 2 3 4 5 6 7 8 9 10 Very

Why did you eat this snack?

Lunch

	Amount	Calories	Time:	Location:
Total Lunch Calories:				

How hungry were you before eating?
Not at all 1 2 3 4 5 6 7 8 9 10 Very

How do you feel about the healthiness and size of this meal?

large/ unhealthy   small/ healthy

☐ ☐ ☐ ☐ ☐

Well Done!

Snack

	Amount	Calories	Time:	Location:
Total Snack Calories:				

How hungry were you before eating?
Not at all 1 2 3 4 5 6 7 8 9 10 Very

Why did you eat this snack?

				Weight	
				Time	

D i n n e r

	Amount	Calories	Time:	Location:

How hungry were you before eating?
Not at all 1 2 3 4 5 6 7 8 9 10 **Very**

How do you feel about the healthiness and size of this meal?

large/
unhealthy

small/
healthy

☐ ☐ ☐ ☐ ☐

Total Dinner Calories:

Well Done!

S n a c k

	Amount	Calories	Time:	Location:

How hungry were you before eating?
Not at all 1 2 3 4 5 6 7 8 9 10 **Very**

Why did you eat this snack?

Total Snack Calories:

Total Daily Calories:

E x e r c i s e	Target	Achievement

Are you happy with how you ate and exercised today?

F o o d	E x e r c i s e
☐ ☐ ☐ ☐ ☐	☐ ☐ ☐ ☐ ☐
Well Done!	**Well Done!**

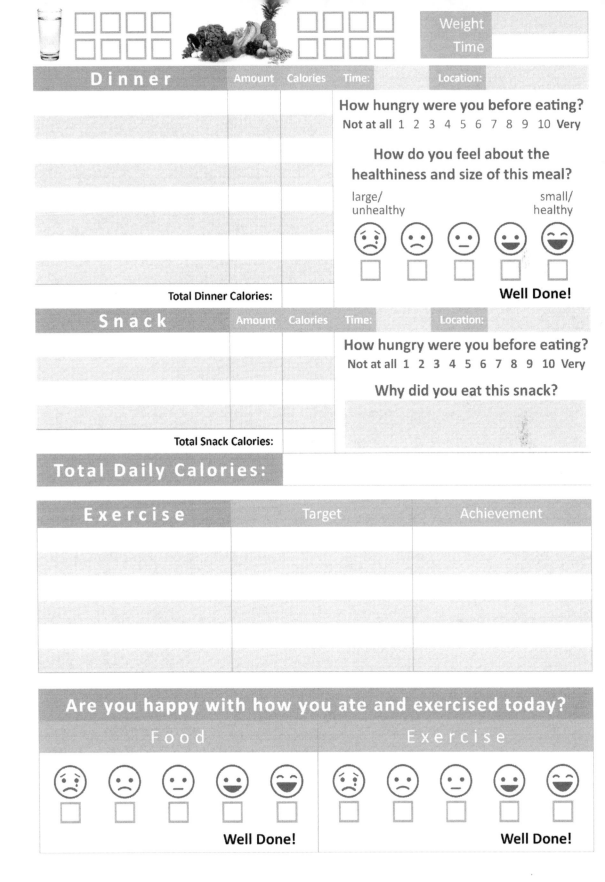

Tuesday Week 2

Month   Date

Breakfast	Amount	Calories	Time:	Location:
Total Breakfast Calories:				

How hungry were you before eating?

Not at all 1 2 3 4 5 6 7 8 9 10 Very

How do you feel about the healthiness and size of this meal?

large/unhealthy   small/healthy

☐ ☐ ☐ ☐ ☐

Well Done!

Snack	Amount	Calories	Time:	Location:
Total Snack Calories:				

How hungry were you before eating?

Not at all 1 2 3 4 5 6 7 8 9 10 Very

Why did you eat this snack?

Lunch	Amount	Calories	Time:	Location:
Total Lunch Calories:				

How hungry were you before eating?

Not at all 1 2 3 4 5 6 7 8 9 10 Very

How do you feel about the healthiness and size of this meal?

large/unhealthy   small/healthy

☐ ☐ ☐ ☐ ☐

Well Done!

Snack	Amount	Calories	Time:	Location:
Total Snack Calories:				

How hungry were you before eating?

Not at all 1 2 3 4 5 6 7 8 9 10 Very

Why did you eat this snack?

Dinner	Amount	Calories	Time:	Location:
Total Dinner Calories:				

How hungry were you before eating?

Not at all 1 2 3 4 5 6 7 8 9 10 Very

How do you feel about the healthiness and size of this meal?

large/unhealthy   small/healthy

☹ ☹ 😐 🙂 😄

☐ ☐ ☐ ☐ ☐

Well Done!

Snack	Amount	Calories	Time:	Location:
Total Snack Calories:				

How hungry were you before eating?

Not at all 1 2 3 4 5 6 7 8 9 10 Very

Why did you eat this snack?

Total Daily Calories:

Exercise	Target	Achievement

Are you happy with how you ate and exercised today?

Food	Exercise
☹ ☹ 😐 🙂 😄	☹ ☹ 😐 🙂 😄
☐ ☐ ☐ ☐ ☐	☐ ☐ ☐ ☐ ☐
Well Done!	**Well Done!**

Weight

Time

Wednesday Week 2

Breakfast	Amount	Calories	Time:	Location:
Total Breakfast Calories:				

How hungry were you before eating?
Not at all 1 2 3 4 5 6 7 8 9 10 Very

How do you feel about the healthiness and size of this meal?

large/unhealthy   small/healthy

☹ ☹ 😐 🙂 😄
☐ ☐ ☐ ☐ ☐

Well Done!

Snack	Amount	Calories	Time:	Location:
Total Snack Calories:				

How hungry were you before eating?
Not at all 1 2 3 4 5 6 7 8 9 10 Very

Why did you eat this snack?

Lunch	Amount	Calories	Time:	Location:
Total Lunch Calories:				

How hungry were you before eating?
Not at all 1 2 3 4 5 6 7 8 9 10 Very

How do you feel about the healthiness and size of this meal?

large/unhealthy   small/healthy

☹ ☹ 😐 🙂 😄
☐ ☐ ☐ ☐ ☐

Well Done!

Snack	Amount	Calories	Time:	Location:
Total Snack Calories:				

How hungry were you before eating?
Not at all 1 2 3 4 5 6 7 8 9 10 Very

Why did you eat this snack?

Weight	
Time	

D i n n e r

	Amount	Calories	Time:	Location:
Total Dinner Calories:				

How hungry were you before eating?
Not at all 1 2 3 4 5 6 7 8 9 10 **Very**

How do you feel about the healthiness and size of this meal?

large/
unhealthy   small/
healthy

☹ 😟 😐 😃 😄
□ □ □ □ □

Well Done!

S n a c k

	Amount	Calories	Time:	Location:
Total Snack Calories:				

How hungry were you before eating?
Not at all 1 2 3 4 5 6 7 8 9 10 **Very**

Why did you eat this snack?

Total Daily Calories:

Exercise	Target	Achievement

Are you happy with how you ate and exercised today?

Food	Exercise

☹ 😟 😐 😃 😄   ☹ 😟 😐 😃 😄
□ □ □ □ □   □ □ □ □ □

Well Done!   **Well Done!**

Thursday Week 2

Breakfast | Amount | Calories | Time: | Location:

How hungry were you before eating?
Not at all 1 2 3 4 5 6 7 8 9 10 Very

How do you feel about the healthiness and size of this meal?

large/unhealthy   small/healthy

☐   ☐   ☐   ☐   ☐

Well Done!

Total Breakfast Calories:

Snack | Amount | Calories | Time: | Location:

How hungry were you before eating?
Not at all 1 2 3 4 5 6 7 8 9 10 Very

Why did you eat this snack?

Total Snack Calories:

Lunch | Amount | Calories | Time: | Location:

How hungry were you before eating?
Not at all 1 2 3 4 5 6 7 8 9 10 Very

How do you feel about the healthiness and size of this meal?

large/unhealthy   small/healthy

☐   ☐   ☐   ☐   ☐

Well Done!

Total Lunch Calories:

Snack | Amount | Calories | Time: | Location:

How hungry were you before eating?
Not at all 1 2 3 4 5 6 7 8 9 10 Very

Why did you eat this snack?

Total Snack Calories:

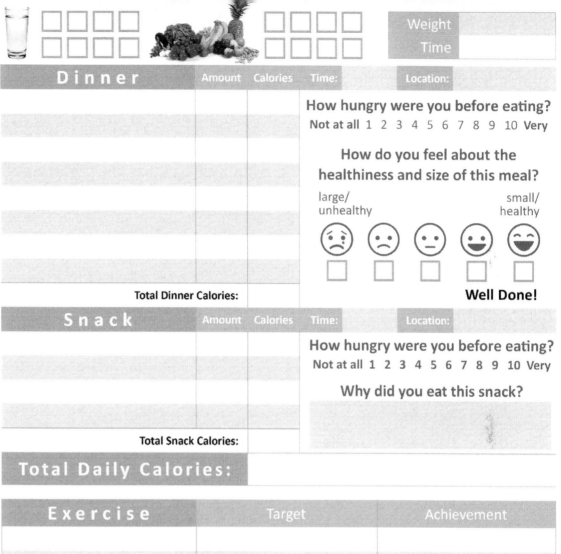

			Weight	
			Time	

Dinner

	Amount	Calories	Time:	Location:
Total Dinner Calories:				

How hungry were you before eating?
Not at all 1 2 3 4 5 6 7 8 9 10 Very

How do you feel about the healthiness and size of this meal?

large/unhealthy   small/healthy

☐ ☐ ☐ ☐ ☐

Well Done!

Snack

	Amount	Calories	Time:	Location:
Total Snack Calories:				

How hungry were you before eating?
Not at all 1 2 3 4 5 6 7 8 9 10 Very

Why did you eat this snack?

Total Daily Calories:

Exercise	Target	Achievement

Are you happy with how you ate and exercised today?

Food	Exercise

☹ ☹ 😐 🙂 😄   ☹ ☹ 😐 🙂 😄
☐ ☐ ☐ ☐ ☐   ☐ ☐ ☐ ☐ ☐

Well Done!   **Well Done!**

Friday Week 2

Breakfast	Amount	Calories	Time:	Location:
Total Breakfast Calories:				

How hungry were you before eating?
Not at all 1 2 3 4 5 6 7 8 9 10 Very

How do you feel about the healthiness and size of this meal?

large/unhealthy   small/healthy

☐ ☐ ☐ ☐ ☐

Well Done!

Snack	Amount	Calories	Time:	Location:
Total Snack Calories:				

How hungry were you before eating?
Not at all 1 2 3 4 5 6 7 8 9 10 Very

Why did you eat this snack?

Lunch	Amount	Calories	Time:	Location:
Total Lunch Calories:				

How hungry were you before eating?
Not at all 1 2 3 4 5 6 7 8 9 10 Very

How do you feel about the healthiness and size of this meal?

large/unhealthy   small/healthy

☐ ☐ ☐ ☐ ☐

Well Done!

Snack	Amount	Calories	Time:	Location:
Total Snack Calories:				

How hungry were you before eating?
Not at all 1 2 3 4 5 6 7 8 9 10 Very

Why did you eat this snack?

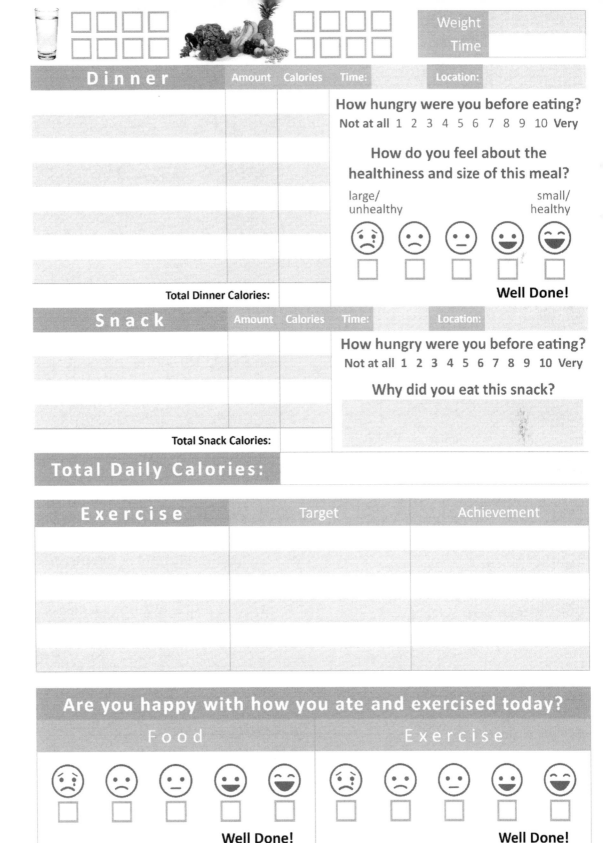

Dinner	Amount	Calories	Time:	Location:
Total Dinner Calories:				

How hungry were you before eating?
Not at all 1 2 3 4 5 6 7 8 9 10 Very

How do you feel about the healthiness and size of this meal?

large/unhealthy   small/healthy

☐ ☐ ☐ ☐ ☐

Well Done!

Snack	Amount	Calories	Time:	Location:
Total Snack Calories:				

How hungry were you before eating?
Not at all 1 2 3 4 5 6 7 8 9 10 Very

Why did you eat this snack?

Total Daily Calories:

Exercise	Target	Achievement

Are you happy with how you ate and exercised today?

Food	Exercise
☐ ☐ ☐ ☐ ☐	☐ ☐ ☐ ☐ ☐
Well Done!	**Well Done!**

Weight

Time

Saturday Week 2

Breakfast	Amount	Calories	Time:	Location:
Total Breakfast Calories:				

How hungry were you before eating?
Not at all 1 2 3 4 5 6 7 8 9 10 Very

How do you feel about the healthiness and size of this meal?

large/unhealthy · small/healthy

☹ ☹ 😐 😊 😄

☐ ☐ ☐ ☐ ☐

Well Done!

Snack	Amount	Calories	Time:	Location:
Total Snack Calories:				

How hungry were you before eating?
Not at all 1 2 3 4 5 6 7 8 9 10 Very

Why did you eat this snack?

Lunch	Amount	Calories	Time:	Location:
Total Lunch Calories:				

How hungry were you before eating?
Not at all 1 2 3 4 5 6 7 8 9 10 Very

How do you feel about the healthiness and size of this meal?

large/unhealthy · small/healthy

☹ ☹ 😐 😊 😄

☐ ☐ ☐ ☐ ☐

Well Done!

Snack	Amount	Calories	Time:	Location:
Total Snack Calories:				

How hungry were you before eating?
Not at all 1 2 3 4 5 6 7 8 9 10 Very

Why did you eat this snack?

Weight

Time

D i n n e r

	Amount	Calories	Time:	Location:
Total Dinner Calories:				

How hungry were you before eating?

Not at all 1 2 3 4 5 6 7 8 9 10 **Very**

How do you feel about the healthiness and size of this meal?

large/
unhealthy

small/
healthy

☹️ ☹️ 😐 😃 😄

Well Done!

S n a c k

	Amount	Calories	Time:	Location:
Total Snack Calories:				

How hungry were you before eating?

Not at all 1 2 3 4 5 6 7 8 9 10 **Very**

Why did you eat this snack?

Total Daily Calories:

E x e r c i s e	Target	Achievement

Are you happy with how you ate and exercised today?

Food

☹️ ☹️ 😐 😃 😄

Well Done!

Exercise

☹️ ☹️ 😐 😃 😄

Well Done!

Sunday Week 2

Breakfast

	Amount	Calories	Time:	Location:
Total Breakfast Calories:				

How hungry were you before eating?

Not at all 1 2 3 4 5 6 7 8 9 10 Very

How do you feel about the healthiness and size of this meal?

large/unhealthy   small/healthy

☹ 😦 😐 🙂 😄

☐ ☐ ☐ ☐ ☐

Well Done!

Snack

	Amount	Calories	Time:	Location:
Total Snack Calories:				

How hungry were you before eating?

Not at all 1 2 3 4 5 6 7 8 9 10 Very

Why did you eat this snack?

Lunch

	Amount	Calories	Time:	Location:
Total Lunch Calories:				

How hungry were you before eating?

Not at all 1 2 3 4 5 6 7 8 9 10 Very

How do you feel about the healthiness and size of this meal?

large/unhealthy   small/healthy

☹ 😦 😐 🙂 😄

☐ ☐ ☐ ☐ ☐

Well Done!

Snack

	Amount	Calories	Time:	Location:
Total Snack Calories:				

How hungry were you before eating?

Not at all 1 2 3 4 5 6 7 8 9 10 Very

Why did you eat this snack?

	Weight	
	Time	

Dinner

	Amount	Calories	Time:	Location:

How hungry were you before eating?
Not at all 1 2 3 4 5 6 7 8 9 10 Very

How do you feel about the healthiness and size of this meal?

large/unhealthy   small/healthy

☹ ☹ 😐 😃 😄

☐ ☐ ☐ ☐ ☐

Total Dinner Calories: **Well Done!**

Snack

	Amount	Calories	Time:	Location:

How hungry were you before eating?
Not at all 1 2 3 4 5 6 7 8 9 10 Very

Why did you eat this snack?

Total Snack Calories:

Total Daily Calories:

Exercise	Target	Achievement

Are you happy with how you ate and exercised today?

Food	Exercise
☹ ☹ 😐 😃 😄	☹ ☹ 😐 😃 😄
☐ ☐ ☐ ☐ ☐	☐ ☐ ☐ ☐ ☐
Well Done!	**Well Done!**

Your Weekly Progress

Date: _____

	This week's measurements
Weight:	
Chest:	
Waist:	
Hips:	
Thighs:	
Calves:	
Upper arms:	
Cholesterol:	
Blood pressure:	

How do you feel about this week's progress?

Well Done!

Things you did well this week:

Things you can improve:

Week

3

Month   Date

Breakfast

	Amount	Calories	Time:	Location:

How hungry were you before eating?

Not at all 1 2 3 4 5 6 7 8 9 10 Very

How do you feel about the healthiness and size of this meal?

large/
unhealthy   small/
healthy

☹ ☹ 😐 😊 😄
☐   ☐   ☐   ☐   ☐

Well Done!

Total Breakfast Calories:

Snack

	Amount	Calories	Time:	Location:

How hungry were you before eating?

Not at all 1 2 3 4 5 6 7 8 9 10 Very

Why did you eat this snack?

Total Snack Calories:

Lunch

	Amount	Calories	Time:	Location:

How hungry were you before eating?

Not at all 1 2 3 4 5 6 7 8 9 10 Very

How do you feel about the healthiness and size of this meal?

large/
unhealthy   small/
healthy

☹ ☹ 😐 😊 😄
☐   ☐   ☐   ☐   ☐

Well Done!

Total Lunch Calories:

Snack

	Amount	Calories	Time:	Location:

How hungry were you before eating?

Not at all 1 2 3 4 5 6 7 8 9 10 Very

Why did you eat this snack?

Total Snack Calories:

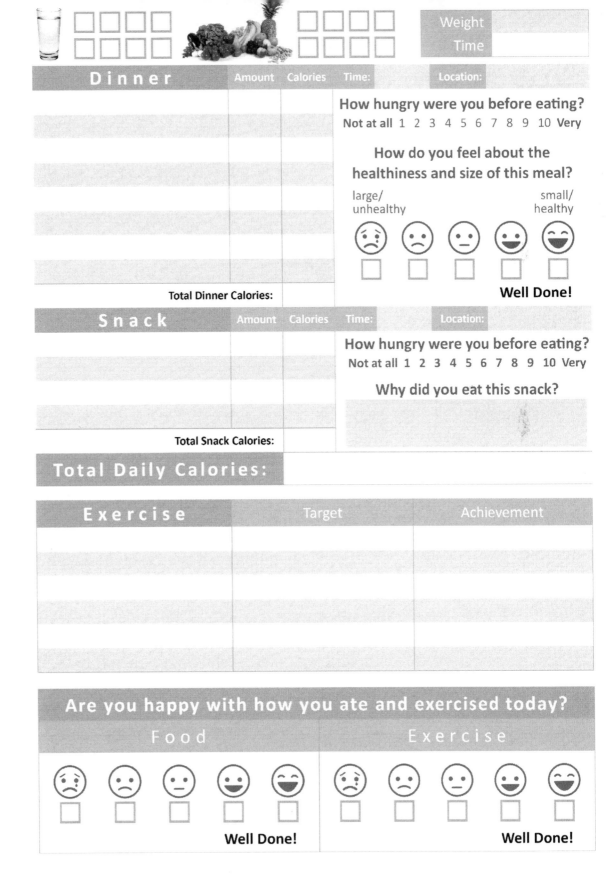

| | Weight |
| | Time |

Dinner

	Amount	Calories	Time:	Location:
Total Dinner Calories:				

How hungry were you before eating?
Not at all 1 2 3 4 5 6 7 8 9 10 Very

How do you feel about the healthiness and size of this meal?

large/unhealthy   small/healthy

☐ ☐ ☐ ☐ ☐

Well Done!

Snack

	Amount	Calories	Time:	Location:
Total Snack Calories:				

How hungry were you before eating?
Not at all 1 2 3 4 5 6 7 8 9 10 Very

Why did you eat this snack?

Total Daily Calories:

Exercise	Target	Achievement

Are you happy with how you ate and exercised today?

Food	Exercise
☐ ☐ ☐ ☐ ☐	☐ ☐ ☐ ☐ ☐
Well Done!	**Well Done!**

Tuesday Week 3

Month   Date

Breakfast	Amount	Calories	Time:	Location:
Total Breakfast Calories:				

How hungry were you before eating?
Not at all 1 2 3 4 5 6 7 8 9 10 Very

How do you feel about the healthiness and size of this meal?

large/unhealthy   small/healthy

☹ ☐ 🙁 ☐ 😐 ☐ 🙂 ☐ 😄 ☐

Well Done!

Snack	Amount	Calories	Time:	Location:
Total Snack Calories:				

How hungry were you before eating?
Not at all 1 2 3 4 5 6 7 8 9 10 Very

Why did you eat this snack?

Lunch	Amount	Calories	Time:	Location:
Total Lunch Calories:				

How hungry were you before eating?
Not at all 1 2 3 4 5 6 7 8 9 10 Very

How do you feel about the healthiness and size of this meal?

large/unhealthy   small/healthy

☹ ☐ 🙁 ☐ 😐 ☐ 🙂 ☐ 😄 ☐

Well Done!

Snack	Amount	Calories	Time:	Location:
Total Snack Calories:				

How hungry were you before eating?
Not at all 1 2 3 4 5 6 7 8 9 10 Very

Why did you eat this snack?

Dinner	Amount	Calories	Time:	Location:
Total Dinner Calories:				

How hungry were you before eating?
Not at all 1 2 3 4 5 6 7 8 9 10 Very

How do you feel about the healthiness and size of this meal?

large/unhealthy   small/healthy

Well Done!

Snack	Amount	Calories	Time:	Location:
Total Snack Calories:				

How hungry were you before eating?
Not at all 1 2 3 4 5 6 7 8 9 10 Very

Why did you eat this snack?

Total Daily Calories:

Exercise	Target	Achievement

Are you happy with how you ate and exercised today?

Food	Exercise
Well Done!	**Well Done!**

Wednesday Week 3

Breakfast	Amount	Calories	Time:	Location:
Total Breakfast Calories:				

How hungry were you before eating?

Not at all 1 2 3 4 5 6 7 8 9 10 Very

How do you feel about the healthiness and size of this meal?

large/ unhealthy   small/ healthy

☹ ☹ 😐 🙂 😄

☐ ☐ ☐ ☐ ☐

Well Done!

Snack	Amount	Calories	Time:	Location:
Total Snack Calories:				

How hungry were you before eating?

Not at all 1 2 3 4 5 6 7 8 9 10 Very

Why did you eat this snack?

Lunch	Amount	Calories	Time:	Location:
Total Lunch Calories:				

How hungry were you before eating?

Not at all 1 2 3 4 5 6 7 8 9 10 Very

How do you feel about the healthiness and size of this meal?

large/ unhealthy   small/ healthy

☹ ☹ 😐 🙂 😄

☐ ☐ ☐ ☐ ☐

Well Done!

Snack	Amount	Calories	Time:	Location:
Total Snack Calories:				

How hungry were you before eating?

Not at all 1 2 3 4 5 6 7 8 9 10 Very

Why did you eat this snack?

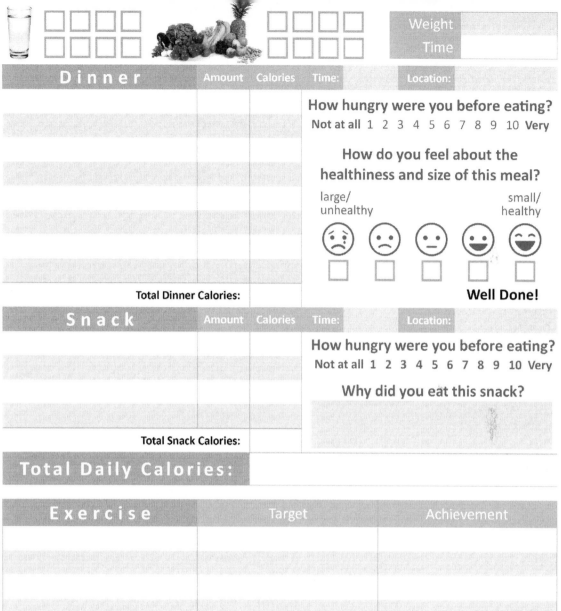

Dinner	Amount	Calories	Time:	Location:
Total Dinner Calories:				

How hungry were you before eating?

Not at all 1 2 3 4 5 6 7 8 9 10 Very

How do you feel about the healthiness and size of this meal?

large/unhealthy   small/healthy

Well Done!

Snack	Amount	Calories	Time:	Location:
Total Snack Calories:				

How hungry were you before eating?

Not at all 1 2 3 4 5 6 7 8 9 10 Very

Why did you eat this snack?

Total Daily Calories:

Exercise	Target	Achievement

Are you happy with how you ate and exercised today?

Food

Exercise

Well Done!   **Well Done!**

Thursday Week 3

B r e a k f a s t	Amount	Calories	Time:	Location:

Total Breakfast Calories:

How hungry were you before eating?

Not at all   1   2   3   4   5   6   7   8   9   10   Very

How do you feel about the healthiness and size of this meal?

large/ unhealthy   small/ healthy

☐   ☐   ☐   ☐   ☐

Well Done!

S n a c k	Amount	Calories	Time:	Location:

Total Snack Calories:

How hungry were you before eating?

Not at all   1   2   3   4   5   6   7   8   9   10   Very

Why did you eat this snack?

L u n c h	Amount	Calories	Time:	Location:

Total Lunch Calories:

How hungry were you before eating?

Not at all   1   2   3   4   5   6   7   8   9   10   Very

How do you feel about the healthiness and size of this meal?

large/ unhealthy   small/ healthy

☐   ☐   ☐   ☐   ☐

Well Done!

S n a c k	Amount	Calories	Time:	Location:

Total Snack Calories:

How hungry were you before eating?

Not at all   1   2   3   4   5   6   7   8   9   10   Very

Why did you eat this snack?

		Weight
		Time

Dinner

	Amount	Calories	Time:	Location:
Total Dinner Calories:				

How hungry were you before eating?
Not at all 1 2 3 4 5 6 7 8 9 10 Very

How do you feel about the healthiness and size of this meal?

large/unhealthy   small/healthy

☹ 😕 😐 😃 😄

☐ ☐ ☐ ☐ ☐

Well Done!

Snack

	Amount	Calories	Time:	Location:
Total Snack Calories:				

How hungry were you before eating?
Not at all 1 2 3 4 5 6 7 8 9 10 Very

Why did you eat this snack?

Total Daily Calories:

Exercise	Target	Achievement

Are you happy with how you ate and exercised today?

Food

☹ 😕 😐 😃 😄

☐ ☐ ☐ ☐ ☐

Well Done!

Exercise

☹ 😕 😐 😃 😄

☐ ☐ ☐ ☐ ☐

Well Done!

Friday Week 3

Breakfast

	Amount	Calories	Time:	Location:

Total Breakfast Calories:

How hungry were you before eating?

Not at all 1 2 3 4 5 6 7 8 9 10 Very

How do you feel about the healthiness and size of this meal?

large/ unhealthy   small/ healthy

☐   ☐   ☐   ☐   ☐

Well Done!

Snack

	Amount	Calories	Time:	Location:

Total Snack Calories:

How hungry were you before eating?

Not at all 1 2 3 4 5 6 7 8 9 10 Very

Why did you eat this snack?

Lunch

	Amount	Calories	Time:	Location:

Total Lunch Calories:

How hungry were you before eating?

Not at all 1 2 3 4 5 6 7 8 9 10 Very

How do you feel about the healthiness and size of this meal?

large/ unhealthy   small/ healthy

☐   ☐   ☐   ☐   ☐

Well Done!

Snack

	Amount	Calories	Time:	Location:

Total Snack Calories:

How hungry were you before eating?

Not at all 1 2 3 4 5 6 7 8 9 10 Very

Why did you eat this snack?

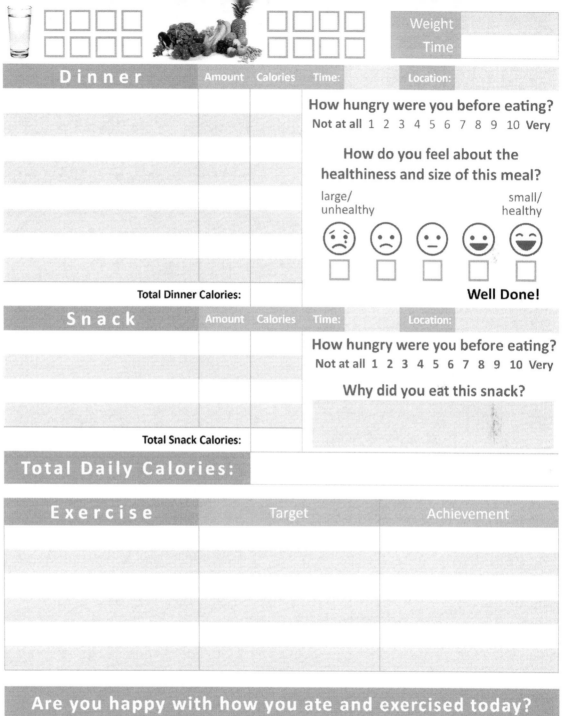

Weight	
Time	

D i n n e r
	Amount	Calories	Time:	Location:

How hungry were you before eating?
Not at all 1 2 3 4 5 6 7 8 9 10 Very

How do you feel about the healthiness and size of this meal?

large/
unhealthy   small/
healthy

☐ ☐ ☐ ☐ ☐

Well Done!

Total Dinner Calories:

S n a c k
	Amount	Calories	Time:	Location:

How hungry were you before eating?
Not at all 1 2 3 4 5 6 7 8 9 10 Very

Why did you eat this snack?

Total Snack Calories:

Total Daily Calories:

E x e r c i s e
	Target	Achievement

Are you happy with how you ate and exercised today?

F o o d

☐ ☐ ☐ ☐ ☐

Well Done!

E x e r c i s e

☐ ☐ ☐ ☐ ☐

Well Done!

Saturday Week 3

Breakfast	Amount	Calories	Time:	Location:
Total Breakfast Calories:				

How hungry were you before eating?
Not at all 1 2 3 4 5 6 7 8 9 10 Very

How do you feel about the healthiness and size of this meal?

large/unhealthy   small/healthy

☹ 🙁 😐 😃 😄
☐   ☐   ☐   ☐   ☐

Well Done!

Snack	Amount	Calories	Time:	Location:
Total Snack Calories:				

How hungry were you before eating?
Not at all 1 2 3 4 5 6 7 8 9 10 Very

Why did you eat this snack?

Lunch	Amount	Calories	Time:	Location:
Total Lunch Calories:				

How hungry were you before eating?
Not at all 1 2 3 4 5 6 7 8 9 10 Very

How do you feel about the healthiness and size of this meal?

large/unhealthy   small/healthy

☹ 🙁 😐 😃 😄
☐   ☐   ☐   ☐   ☐

Well Done!

Snack	Amount	Calories	Time:	Location:
Total Snack Calories:				

How hungry were you before eating?
Not at all 1 2 3 4 5 6 7 8 9 10 Very

Why did you eat this snack?

| | | Weight |
| | | Time |

Dinner

	Amount	Calories	Time:	Location:
Total Dinner Calories:				

How hungry were you before eating?

Not at all 1 2 3 4 5 6 7 8 9 10 **Very**

How do you feel about the healthiness and size of this meal?

large/unhealthy   small/healthy

☹ 🙁 😐 😀 😄

☐ ☐ ☐ ☐ ☐

Well Done!

Snack

	Amount	Calories	Time:	Location:
Total Snack Calories:				

How hungry were you before eating?

Not at all 1 2 3 4 5 6 7 8 9 10 **Very**

Why did you eat this snack?

Total Daily Calories:

Exercise	Target	Achievement

Are you happy with how you ate and exercised today?

Food	Exercise
☹ 🙁 😐 😀 😄	☹ 🙁 😐 😀 😄
☐ ☐ ☐ ☐ ☐	☐ ☐ ☐ ☐ ☐
Well Done!	**Well Done!**

Sunday Week 3

Month	Date

Breakfast

	Amount	Calories	Time:	Location:

Total Breakfast Calories:

How hungry were you before eating?
Not at all 1 2 3 4 5 6 7 8 9 10 Very

How do you feel about the healthiness and size of this meal?

large/unhealthy   small/healthy

☐ ☐ ☐ ☐ ☐

Well Done!

Snack

	Amount	Calories	Time:	Location:

Total Snack Calories:

How hungry were you before eating?
Not at all 1 2 3 4 5 6 7 8 9 10 Very

Why did you eat this snack?

Lunch

	Amount	Calories	Time:	Location:

Total Lunch Calories:

How hungry were you before eating?
Not at all 1 2 3 4 5 6 7 8 9 10 Very

How do you feel about the healthiness and size of this meal?

large/unhealthy   small/healthy

☐ ☐ ☐ ☐ ☐

Well Done!

Snack

	Amount	Calories	Time:	Location:

Total Snack Calories:

How hungry were you before eating?
Not at all 1 2 3 4 5 6 7 8 9 10 Very

Why did you eat this snack?

				Weight	
				Time	

D i n n e r

	Amount	Calories	Time:	Location:

Total Dinner Calories:

How hungry were you before eating?
Not at all 1 2 3 4 5 6 7 8 9 10 **Very**

How do you feel about the healthiness and size of this meal?

large/ unhealthy   small/ healthy

☹ ☹ 😐 🙂 😄
☐ ☐ ☐ ☐ ☐

Well Done!

S n a c k

	Amount	Calories	Time:	Location:

Total Snack Calories:

How hungry were you before eating?
Not at all 1 2 3 4 5 6 7 8 9 10 **Very**

Why did you eat this snack?

Total Daily Calories:

E x e r c i s e	Target	Achievement

Are you happy with how you ate and exercised today?

F o o d	E x e r c i s e

☹ ☹ 😐 🙂 😄   ☹ ☹ 😐 🙂 😄
☐ ☐ ☐ ☐ ☐   ☐ ☐ ☐ ☐ ☐

Well Done!   **Well Done!**

Your Weekly Progress

Date: _____

	This week's measurements
Weight:	
Chest:	
Waist:	
Hips:	
Thighs:	
Calves:	
Upper arms:	
Cholesterol:	
Blood pressure:	

How do you feel about this week's progress?

Well Done!

Things you did well this week:

Things you can improve:

Week 4

MondayWeek 4

B r e a k f a s t	Amount	Calories	Time:	Location:
Total Breakfast Calories:				

How hungry were you before eating?

Not at all 1 2 3 4 5 6 7 8 9 10 Very

How do you feel about the healthiness and size of this meal?

large/
unhealthy   small/
                                                    healthy

☹ ☹ 😐 🙂 😄

☐ ☐ ☐ ☐ ☐

Well Done!

S n a c k	Amount	Calories	Time:	Location:
Total Snack Calories:				

How hungry were you before eating?

Not at all 1 2 3 4 5 6 7 8 9 10 Very

Why did you eat this snack?

L u n c h	Amount	Calories	Time:	Location:
Total Lunch Calories:				

How hungry were you before eating?

Not at all 1 2 3 4 5 6 7 8 9 10 Very

How do you feel about the healthiness and size of this meal?

large/
unhealthy   small/
                                                    healthy

☹ ☹ 😐 🙂 😄

☐ ☐ ☐ ☐ ☐

Well Done!

S n a c k	Amount	Calories	Time:	Location:
Total Snack Calories:				

How hungry were you before eating?

Not at all 1 2 3 4 5 6 7 8 9 10 Very

Why did you eat this snack?

				Weight
				Time

Dinner

	Amount	Calories	Time:	Location:

How hungry were you before eating?

Not at all 1 2 3 4 5 6 7 8 9 10 **Very**

How do you feel about the healthiness and size of this meal?

large/
unhealthy

small/
healthy

☹ 🙁 😐 😃 😄

☐ ☐ ☐ ☐ ☐

Well Done!

Total Dinner Calories:

Snack

	Amount	Calories	Time:	Location:

How hungry were you before eating?

Not at all 1 2 3 4 5 6 7 8 9 10 **Very**

Why did you eat this snack?

Total Snack Calories:

Total Daily Calories:

Exercise	Target	Achievement

Are you happy with how you ate and exercised today?

Food

☹ 🙁 😐 😃 😄

☐ ☐ ☐ ☐ ☐

Well Done!

Exercise

☹ 🙁 😐 😃 😄

☐ ☐ ☐ ☐ ☐

Well Done!

Tuesday Week 4

Breakfast	Amount	Calories	Time:	Location:
Total Breakfast Calories:				

How hungry were you before eating?
Not at all   1   2   3   4   5   6   7   8   9   10   Very

How do you feel about the healthiness and size of this meal?

large/unhealthy   small/healthy

☹ 😕 😐 😃 😄
☐ ☐ ☐ ☐ ☐

Well Done!

Snack	Amount	Calories	Time:	Location:
Total Snack Calories:				

How hungry were you before eating?
Not at all   1   2   3   4   5   6   7   8   9   10   Very

Why did you eat this snack?

Lunch	Amount	Calories	Time:	Location:
Total Lunch Calories:				

How hungry were you before eating?
Not at all   1   2   3   4   5   6   7   8   9   10   Very

How do you feel about the healthiness and size of this meal?

large/unhealthy   small/healthy

☹ 😕 😐 😃 😄
☐ ☐ ☐ ☐ ☐

Well Done!

Snack	Amount	Calories	Time:	Location:
Total Snack Calories:				

How hungry were you before eating?
Not at all   1   2   3   4   5   6   7   8   9   10   Very

Why did you eat this snack?

| | Weight | |
| | Time | |

D i n n e r

	Amount	Calories	Time:	Location:

How hungry were you before eating?

Not at all 1 2 3 4 5 6 7 8 9 10 **Very**

How do you feel about the healthiness and size of this meal?

large/ unhealthy   small/ healthy

☹ 🙁 😐 😃 😄

☐ ☐ ☐ ☐ ☐

Total Dinner Calories:

Well Done!

S n a c k

	Amount	Calories	Time:	Location:

How hungry were you before eating?

Not at all 1 2 3 4 5 6 7 8 9 10 **Very**

Why did you eat this snack?

Total Snack Calories:

Total Daily Calories:

Exercise	Target	Achievement

Are you happy with how you ate and exercised today?

Food

☹ 🙁 😐 😃 😄

☐ ☐ ☐ ☐ ☐

Well Done!

Exercise

☹ 🙁 😐 😃 😄

☐ ☐ ☐ ☐ ☐

Well Done!

Wednesday Week 4

Breakfast	Amount	Calories	Time:	Location:
Total Breakfast Calories:				

How hungry were you before eating?
Not at all 1 2 3 4 5 6 7 8 9 10 Very

How do you feel about the healthiness and size of this meal?

large/
unhealthy   small/
healthy

☹ ☹ 😐 😀 😄

☐ ☐ ☐ ☐ ☐

Well Done!

Snack	Amount	Calories	Time:	Location:
Total Snack Calories:				

How hungry were you before eating?
Not at all 1 2 3 4 5 6 7 8 9 10 Very

Why did you eat this snack?

Lunch	Amount	Calories	Time:	Location:
Total Lunch Calories:				

How hungry were you before eating?
Not at all 1 2 3 4 5 6 7 8 9 10 Very

How do you feel about the healthiness and size of this meal?

large/
unhealthy   small/
healthy

☹ ☹ 😐 😀 😄

☐ ☐ ☐ ☐ ☐

Well Done!

Snack	Amount	Calories	Time:	Location:
Total Snack Calories:				

How hungry were you before eating?
Not at all 1 2 3 4 5 6 7 8 9 10 Very

Why did you eat this snack?

Weight

Time

D i n n e r	Amount	Calories	Time:	Location:

How hungry were you before eating?

Not at all 1 2 3 4 5 6 7 8 9 10 **Very**

How do you feel about the healthiness and size of this meal?

large/
unhealthy

small/
healthy

☹ ☹ 😐 😊 😄
☐ ☐ ☐ ☐ ☐

Well Done!

Total Dinner Calories:

S n a c k	Amount	Calories	Time:	Location:

How hungry were you before eating?

Not at all 1 2 3 4 5 6 7 8 9 10 **Very**

Why did you eat this snack?

Total Snack Calories:

Total Daily Calories:

E x e r c i s e	Target	Achievement

Are you happy with how you ate and exercised today?

Food	Exercise
☹ ☹ 😐 😊 😄	☹ ☹ 😐 😊 😄
☐ ☐ ☐ ☐ ☐	☐ ☐ ☐ ☐ ☐
Well Done!	**Well Done!**

ThursdayWeek 4

Breakfast	Amount	Calories	Time:	Location:
Total Breakfast Calories:				

How hungry were you before eating?

Not at all   1   2   3   4   5   6   7   8   9   10   Very

How do you feel about the healthiness and size of this meal?

large/ unhealthy   small/ healthy

☐   ☐   ☐   ☐   ☐

Well Done!

Snack	Amount	Calories	Time:	Location:
Total Snack Calories:				

How hungry were you before eating?

Not at all   1   2   3   4   5   6   7   8   9   10   Very

Why did you eat this snack?

Lunch	Amount	Calories	Time:	Location:
Total Lunch Calories:				

How hungry were you before eating?

Not at all   1   2   3   4   5   6   7   8   9   10   Very

How do you feel about the healthiness and size of this meal?

large/ unhealthy   small/ healthy

☐   ☐   ☐   ☐   ☐

Well Done!

Snack	Amount	Calories	Time:	Location:
Total Snack Calories:				

How hungry were you before eating?

Not at all   1   2   3   4   5   6   7   8   9   10   Very

Why did you eat this snack?

	Weight	
	Time	

Dinner

	Amount	Calories	Time:	Location:

How hungry were you before eating?

Not at all 1 2 3 4 5 6 7 8 9 10 Very

How do you feel about the healthiness and size of this meal?

large/unhealthy   small/healthy

☹ 🙁 😐 🙂 😄

☐ ☐ ☐ ☐ ☐

Well Done!

Total Dinner Calories:

Snack

	Amount	Calories	Time:	Location:

How hungry were you before eating?

Not at all 1 2 3 4 5 6 7 8 9 10 Very

Why did you eat this snack?

Total Snack Calories:

Total Daily Calories:

Exercise	Target	Achievement

Are you happy with how you ate and exercised today?

Food	Exercise
☹ 🙁 😐 🙂 😄	☹ 🙁 😐 🙂 😄
☐ ☐ ☐ ☐ ☐	☐ ☐ ☐ ☐ ☐
Well Done!	**Well Done!**

Friday Week 4

Breakfast	Amount	Calories	Time:	Location:

How hungry were you before eating?

Not at all 1 2 3 4 5 6 7 8 9 10 Very

How do you feel about the healthiness and size of this meal?

large/unhealthy   small/healthy

☹ 🙁 😐 🙂 😄

☐ ☐ ☐ ☐ ☐

Total Breakfast Calories: **Well Done!**

Snack	Amount	Calories	Time:	Location:

How hungry were you before eating?

Not at all 1 2 3 4 5 6 7 8 9 10 Very

Why did you eat this snack?

Total Snack Calories:

Lunch	Amount	Calories	Time:	Location:

How hungry were you before eating?

Not at all 1 2 3 4 5 6 7 8 9 10 Very

How do you feel about the healthiness and size of this meal?

large/unhealthy   small/healthy

☹ 🙁 😐 🙂 😄

☐ ☐ ☐ ☐ ☐

Total Lunch Calories: **Well Done!**

Snack	Amount	Calories	Time:	Location:

How hungry were you before eating?

Not at all 1 2 3 4 5 6 7 8 9 10 Very

Why did you eat this snack?

Total Snack Calories:

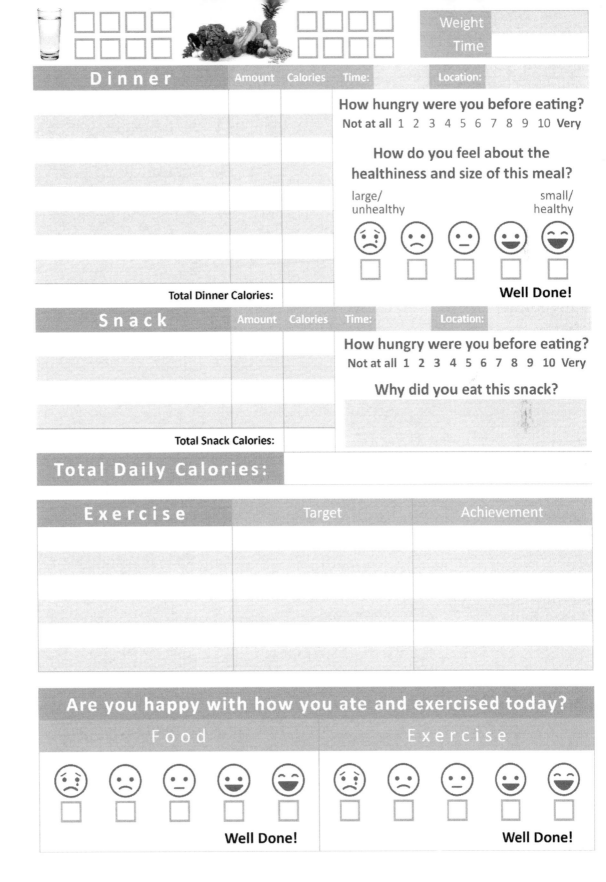

	Weight	
	Time	

Dinner

	Amount	Calories	Time:	Location:

How hungry were you before eating?

Not at all 1 2 3 4 5 6 7 8 9 10 **Very**

How do you feel about the healthiness and size of this meal?

large/unhealthy   small/healthy

☐ ☐ ☐ ☐ ☐

Well Done!

Total Dinner Calories:

Snack

	Amount	Calories	Time:	Location:

How hungry were you before eating?

Not at all 1 2 3 4 5 6 7 8 9 10 **Very**

Why did you eat this snack?

Total Snack Calories:

Total Daily Calories:

Exercise	Target	Achievement

Are you happy with how you ate and exercised today?

Food

☐ ☐ ☐ ☐ ☐

Well Done!

Exercise

☐ ☐ ☐ ☐ ☐

Well Done!

Saturday Week 4

Breakfast

	Amount	Calories	Time:	Location:

Total Breakfast Calories:

How hungry were you before eating?

Not at all 1 2 3 4 5 6 7 8 9 10 Very

How do you feel about the healthiness and size of this meal?

large/ unhealthy   small/ healthy

☐   ☐   ☐   ☐   ☐

Well Done!

Snack

	Amount	Calories	Time:	Location:

Total Snack Calories:

How hungry were you before eating?

Not at all 1 2 3 4 5 6 7 8 9 10 Very

Why did you eat this snack?

Lunch

	Amount	Calories	Time:	Location:

Total Lunch Calories:

How hungry were you before eating?

Not at all 1 2 3 4 5 6 7 8 9 10 Very

How do you feel about the healthiness and size of this meal?

large/ unhealthy   small/ healthy

☐   ☐   ☐   ☐   ☐

Well Done!

Snack

	Amount	Calories	Time:	Location:

Total Snack Calories:

How hungry were you before eating?

Not at all 1 2 3 4 5 6 7 8 9 10 Very

Why did you eat this snack?

							Weight	
							Time	

Dinner	Amount	Calories	Time:	Location:

How hungry were you before eating?

Not at all 1 2 3 4 5 6 7 8 9 10 Very

How do you feel about the healthiness and size of this meal?

large/unhealthy   small/healthy

☐   ☐   ☐   ☐   ☐

Well Done!

Total Dinner Calories:

Snack	Amount	Calories	Time:	Location:

How hungry were you before eating?

Not at all 1 2 3 4 5 6 7 8 9 10 Very

Why did you eat this snack?

Total Snack Calories:

Total Daily Calories:

Exercise	Target	Achievement

Are you happy with how you ate and exercised today?

Food	Exercise

☐   ☐   ☐   ☐   ☐   ☐   ☐   ☐   ☐   ☐

Well Done!   **Well Done!**

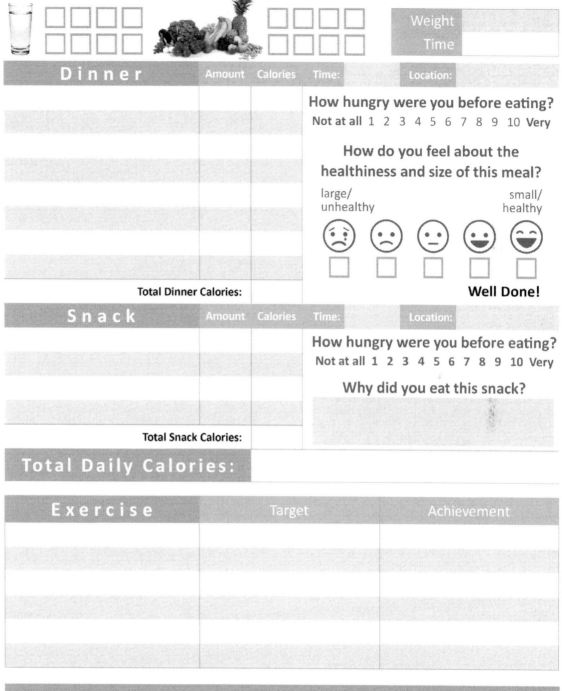

Sunday Week 4

B r e a k f a s t   Amount   Calories   Time: Location:

How hungry were you before eating?

Not at all 1 2 3 4 5 6 7 8 9 10 Very

How do you feel about the healthiness and size of this meal?

large/   small/
unhealthy   healthy

☹️   🙁   😐   😀   😄

☐   ☐   ☐   ☐   ☐

Total Breakfast Calories: **Well Done!**

S n a c k   Amount   Calories   Time: Location:

How hungry were you before eating?

Not at all 1 2 3 4 5 6 7 8 9 10 Very

Why did you eat this snack?

Total Snack Calories:

L u n c h   Amount   Calories   Time: Location:

How hungry were you before eating?

Not at all 1 2 3 4 5 6 7 8 9 10 Very

How do you feel about the healthiness and size of this meal?

large/   small/
unhealthy   healthy

☹️   🙁   😐   😀   😄

☐   ☐   ☐   ☐   ☐

Total Lunch Calories: **Well Done!**

S n a c k   Amount   Calories   Time: Location:

How hungry were you before eating?

Not at all 1 2 3 4 5 6 7 8 9 10 Very

Why did you eat this snack?

Total Snack Calories:

Weight
Time

D i n n e r	Amount	Calories	Time:	Location:
Total Dinner Calories:				

How hungry were you before eating?
Not at all 1 2 3 4 5 6 7 8 9 10 Very

How do you feel about the healthiness and size of this meal?

large/unhealthy   small/healthy

☹	🙁	😐	😃	😄
☐	☐	☐	☐	☐

Well Done!

S n a c k	Amount	Calories	Time:	Location:
Total Snack Calories:				

How hungry were you before eating?
Not at all 1 2 3 4 5 6 7 8 9 10 Very

Why did you eat this snack?

Total Daily Calories:

E x e r c i s e	Target	Achievement

Are you happy with how you ate and exercised today?

F o o d	E x e r c i s e

☹	🙁	😐	😃	😄	☹	🙁	😐	😃	😄
☐	☐	☐	☐	☐	☐	☐	☐	☐	☐

Well Done!   **Well Done!**

Your Weekly Progress

◯ ◯ ◯ ◯ ◯ ◯ ◯ ◯ ◯ ◯ ◯

Date: _____

	This week's measurements
Weight:	
Chest:	
Waist:	
Hips:	
Thighs:	
Calves:	
Upper arms:	
Cholesterol:	
Blood pressure:	

How do you feel about this week's progress?

☐ ☐ ☐ ☐ ☐

Well Done!

Things you did well this week:

Things you can improve:

Week 5

MondayWeek 5

Breakfast	Amount	Calories	Time:	Location:
Total Breakfast Calories:				

How hungry were you before eating?
Not at all 1 2 3 4 5 6 7 8 9 10 Very

How do you feel about the healthiness and size of this meal?

large/unhealthy   small/healthy

☹ 🙁 😐 🙂 😄
☐ ☐ ☐ ☐ ☐

Well Done!

Snack	Amount	Calories	Time:	Location:
Total Snack Calories:				

How hungry were you before eating?
Not at all 1 2 3 4 5 6 7 8 9 10 Very

Why did you eat this snack?

Lunch	Amount	Calories	Time:	Location:
Total Lunch Calories:				

How hungry were you before eating?
Not at all 1 2 3 4 5 6 7 8 9 10 Very

How do you feel about the healthiness and size of this meal?

large/unhealthy   small/healthy

☹ 🙁 😐 🙂 😄
☐ ☐ ☐ ☐ ☐

Well Done!

Snack	Amount	Calories	Time:	Location:
Total Snack Calories:				

How hungry were you before eating?
Not at all 1 2 3 4 5 6 7 8 9 10 Very

Why did you eat this snack?

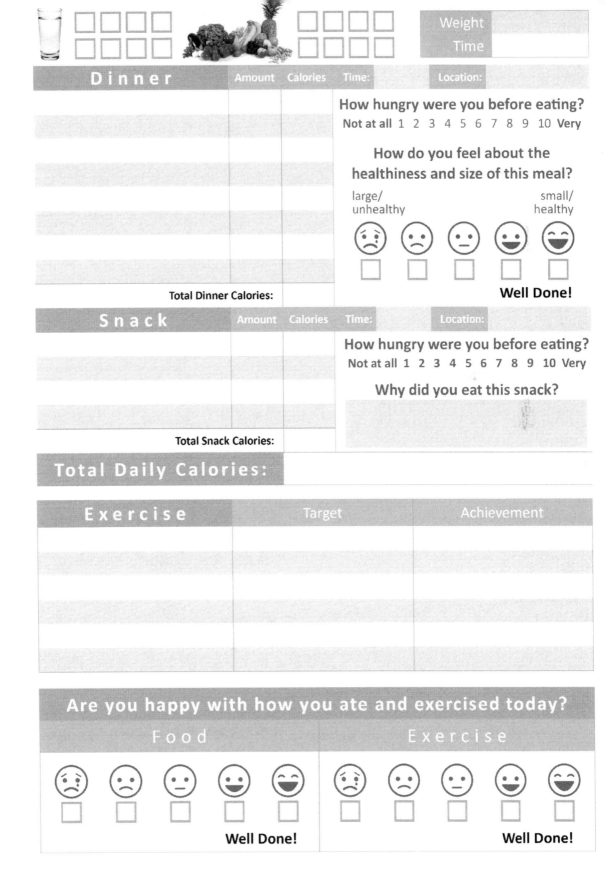

	Weight
	Time

Dinner

	Amount	Calories	Time:	Location:

Total Dinner Calories:

How hungry were you before eating?
Not at all 1 2 3 4 5 6 7 8 9 10 Very

How do you feel about the healthiness and size of this meal?

large/unhealthy   small/healthy

Well Done!

Snack

	Amount	Calories	Time:	Location:

Total Snack Calories:

How hungry were you before eating?
Not at all 1 2 3 4 5 6 7 8 9 10 Very

Why did you eat this snack?

Total Daily Calories:

Exercise	Target	Achievement

Are you happy with how you ate and exercised today?

Food	Exercise

Well Done!   **Well Done!**

Tuesday Week 5

Breakfast

	Amount	Calories	Time:	Location:

How hungry were you before eating?

Not at all 1 2 3 4 5 6 7 8 9 10 Very

How do you feel about the healthiness and size of this meal?

large/unhealthy   small/healthy

☹ 🙁 😐 🙂 😄

☐ ☐ ☐ ☐ ☐

Well Done!

Total Breakfast Calories:

Snack

	Amount	Calories	Time:	Location:

How hungry were you before eating?

Not at all 1 2 3 4 5 6 7 8 9 10 Very

Why did you eat this snack?

Total Snack Calories:

Lunch

	Amount	Calories	Time:	Location:

How hungry were you before eating?

Not at all 1 2 3 4 5 6 7 8 9 10 Very

How do you feel about the healthiness and size of this meal?

large/unhealthy   small/healthy

☹ 🙁 😐 🙂 😄

☐ ☐ ☐ ☐ ☐

Well Done!

Total Lunch Calories:

Snack

	Amount	Calories	Time:	Location:

How hungry were you before eating?

Not at all 1 2 3 4 5 6 7 8 9 10 Very

Why did you eat this snack?

Total Snack Calories:

		Weight	
		Time	

Dinner

	Amount	Calories	Time:	Location:

How hungry were you before eating?
Not at all 1 2 3 4 5 6 7 8 9 10 **Very**

How do you feel about the healthiness and size of this meal?

large/
unhealthy   small/
healthy

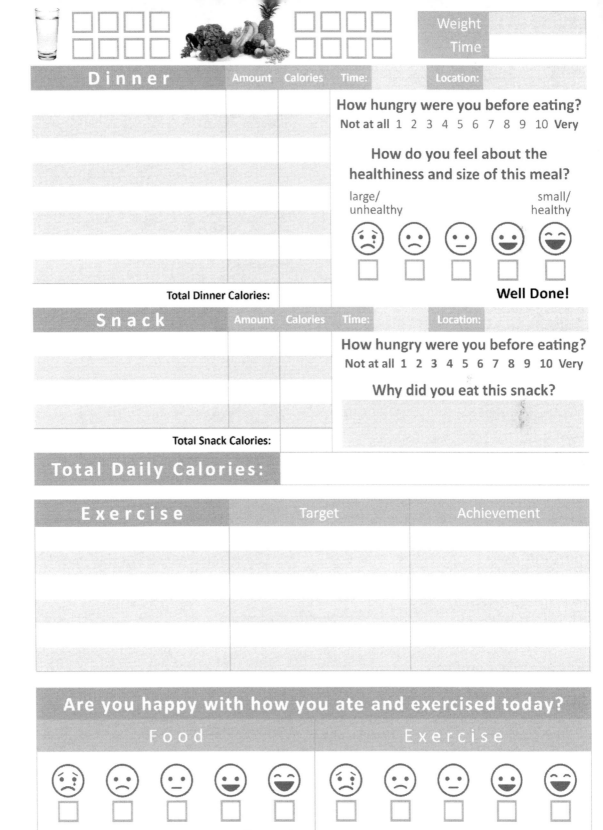

☐ ☐ ☐ ☐ ☐

Well Done!

Total Dinner Calories:

Snack

	Amount	Calories	Time:	Location:

How hungry were you before eating?
Not at all 1 2 3 4 5 6 7 8 9 10 **Very**

Why did you eat this snack?

Total Snack Calories:

Total Daily Calories:

Exercise	Target	Achievement

Are you happy with how you ate and exercised today?

Food	Exercise

☐ ☐ ☐ ☐ ☐   ☐ ☐ ☐ ☐ ☐

Well Done!   **Well Done!**

Wednesday Week 5

Breakfast	Amount	Calories	Time:	Location:

How hungry were you before eating?
Not at all 1 2 3 4 5 6 7 8 9 10 Very

How do you feel about the healthiness and size of this meal?

large/unhealthy   small/healthy

☹ ☐   🙁 ☐   😐 ☐   🙂 ☐   😄 ☐

Well Done!

Total Breakfast Calories:

Snack	Amount	Calories	Time:	Location:

How hungry were you before eating?
Not at all 1 2 3 4 5 6 7 8 9 10 Very

Why did you eat this snack?

Total Snack Calories:

Lunch	Amount	Calories	Time:	Location:

How hungry were you before eating?
Not at all 1 2 3 4 5 6 7 8 9 10 Very

How do you feel about the healthiness and size of this meal?

large/unhealthy   small/healthy

☹ ☐   🙁 ☐   😐 ☐   🙂 ☐   😄 ☐

Well Done!

Total Lunch Calories:

Snack	Amount	Calories	Time:	Location:

How hungry were you before eating?
Not at all 1 2 3 4 5 6 7 8 9 10 Very

Why did you eat this snack?

Total Snack Calories:

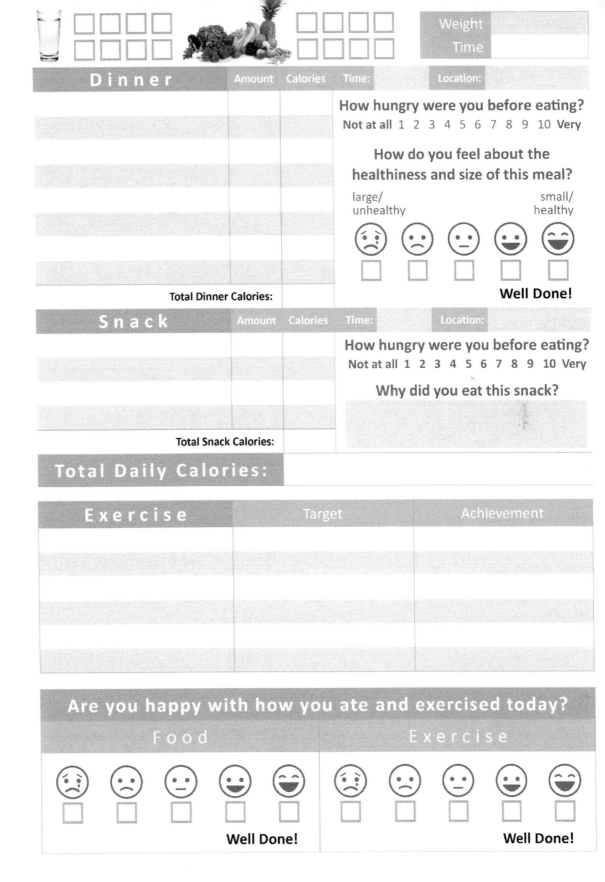

| Weight | |
| Time | |

D i n n e r

	Amount	Calories	Time:	Location:

How hungry were you before eating?

Not at all 1 2 3 4 5 6 7 8 9 10 Very

How do you feel about the healthiness and size of this meal?

large/unhealthy   small/healthy

☹ 🙁 😐 🙂 😄

☐ ☐ ☐ ☐ ☐

Well Done!

Total Dinner Calories:

S n a c k

	Amount	Calories	Time:	Location:

How hungry were you before eating?

Not at all 1 2 3 4 5 6 7 8 9 10 Very

Why did you eat this snack?

Total Snack Calories:

Total Daily Calories:

E x e r c i s e	Target	Achievement

Are you happy with how you ate and exercised today?

F o o d	E x e r c i s e
☹ 🙁 😐 🙂 😄	☹ 🙁 😐 🙂 😄
☐ ☐ ☐ ☐ ☐	☐ ☐ ☐ ☐ ☐
Well Done!	**Well Done!**

Thursday Week 5

Breakfast

	Amount	Calories	Time:	Location:

Total Breakfast Calories:

How hungry were you before eating?
Not at all   1   2   3   4   5   6   7   8   9   10   Very

How do you feel about the healthiness and size of this meal?

large/ unhealthy   small/ healthy

☐   ☐   ☐   ☐   ☐

Well Done!

Snack

	Amount	Calories	Time:	Location:

Total Snack Calories:

How hungry were you before eating?
Not at all   1   2   3   4   5   6   7   8   9   10   Very

Why did you eat this snack?

Lunch

	Amount	Calories	Time:	Location:

Total Lunch Calories:

How hungry were you before eating?
Not at all   1   2   3   4   5   6   7   8   9   10   Very

How do you feel about the healthiness and size of this meal?

large/ unhealthy   small/ healthy

☐   ☐   ☐   ☐   ☐

Well Done!

Snack

	Amount	Calories	Time:	Location:

Total Snack Calories:

How hungry were you before eating?
Not at all   1   2   3   4   5   6   7   8   9   10   Very

Why did you eat this snack?

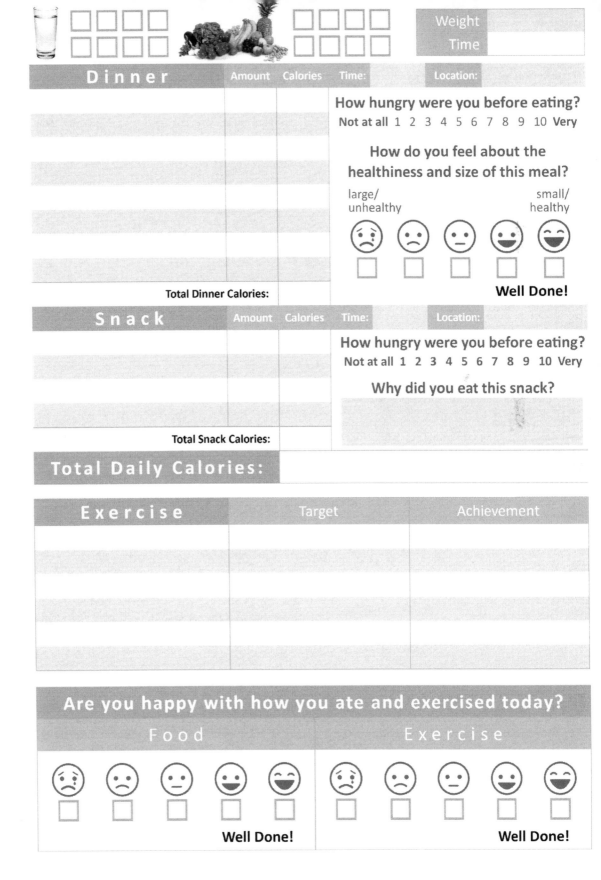

| | Weight |
| | Time |

D i n n e r

	Amount	Calories	Time:	Location:

How hungry were you before eating?
Not at all 1 2 3 4 5 6 7 8 9 10 Very

How do you feel about the healthiness and size of this meal?

large/ unhealthy   small/ healthy

Well Done!

Total Dinner Calories:

S n a c k

	Amount	Calories	Time:	Location:

How hungry were you before eating?
Not at all 1 2 3 4 5 6 7 8 9 10 Very

Why did you eat this snack?

Total Snack Calories:

Total Daily Calories:

Exercise	Target	Achievement

Are you happy with how you ate and exercised today?

Food

Exercise

Well Done!   **Well Done!**

Friday Week 5

Breakfast

Amount	Calories	Time:	Location:

How hungry were you before eating?

Not at all 1 2 3 4 5 6 7 8 9 10 Very

How do you feel about the healthiness and size of this meal?

large/unhealthy   small/healthy

☹ ☹ 😐 🙂 😄
☐ ☐ ☐ ☐ ☐

Well Done!

Total Breakfast Calories:

Snack

Amount	Calories	Time:	Location:

How hungry were you before eating?

Not at all 1 2 3 4 5 6 7 8 9 10 Very

Why did you eat this snack?

Total Snack Calories:

Lunch

Amount	Calories	Time:	Location:

How hungry were you before eating?

Not at all 1 2 3 4 5 6 7 8 9 10 Very

How do you feel about the healthiness and size of this meal?

large/unhealthy   small/healthy

☹ ☹ 😐 🙂 😄
☐ ☐ ☐ ☐ ☐

Well Done!

Total Lunch Calories:

Snack

Amount	Calories	Time:	Location:

How hungry were you before eating?

Not at all 1 2 3 4 5 6 7 8 9 10 Very

Why did you eat this snack?

Total Snack Calories:

Weight	
Time	

D i n n e r

	Amount	Calories	Time:	Location:

How hungry were you before eating?

Not at all 1 2 3 4 5 6 7 8 9 10 Very

How do you feel about the healthiness and size of this meal?

large/unhealthy   small/healthy

☹   ☹   😐   😊   😄

☐   ☐   ☐   ☐   ☐

Well Done!

Total Dinner Calories:

S n a c k

	Amount	Calories	Time:	Location:

How hungry were you before eating?

Not at all 1 2 3 4 5 6 7 8 9 10 Very

Why did you eat this snack?

Total Snack Calories:

Total Daily Calories:

E x e r c i s e	Target	Achievement

Are you happy with how you ate and exercised today?

F o o d

☹   ☹   😐   😊   😄

☐   ☐   ☐   ☐   ☐

Well Done!

E x e r c i s e

☹   ☹   😐   😊   😄

☐   ☐   ☐   ☐   ☐

Well Done!

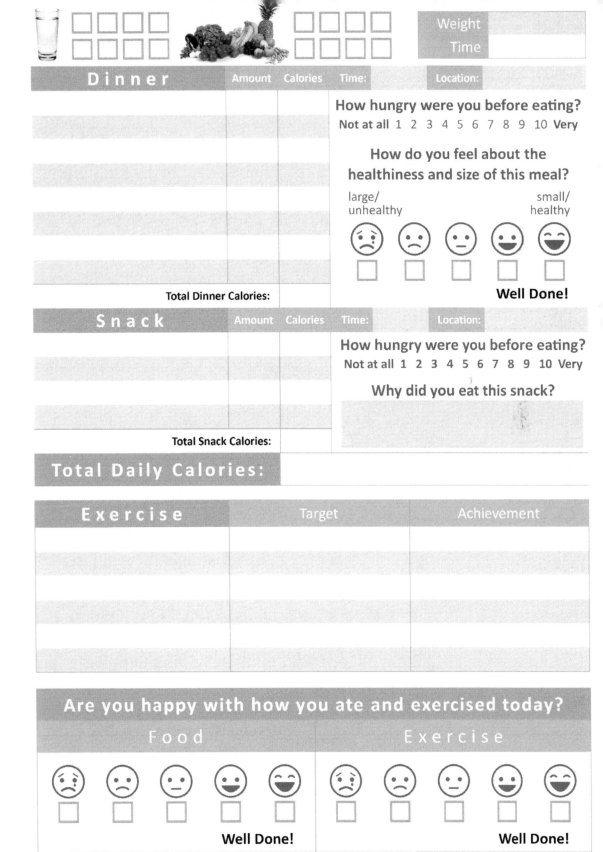

Saturday Week 5

Breakfast

	Amount	Calories	Time:	Location:
Total Breakfast Calories:				

How hungry were you before eating?

Not at all 1 2 3 4 5 6 7 8 9 10 Very

How do you feel about the healthiness and size of this meal?

large/unhealthy   small/healthy

☹ 🙁 😐 🙂 😄
☐ ☐ ☐ ☐ ☐

Well Done!

Snack

	Amount	Calories	Time:	Location:
Total Snack Calories:				

How hungry were you before eating?

Not at all 1 2 3 4 5 6 7 8 9 10 Very

Why did you eat this snack?

Lunch

	Amount	Calories	Time:	Location:
Total Lunch Calories:				

How hungry were you before eating?

Not at all 1 2 3 4 5 6 7 8 9 10 Very

How do you feel about the healthiness and size of this meal?

large/unhealthy   small/healthy

☹ 🙁 😐 🙂 😄
☐ ☐ ☐ ☐ ☐

Well Done!

Snack

	Amount	Calories	Time:	Location:
Total Snack Calories:				

How hungry were you before eating?

Not at all 1 2 3 4 5 6 7 8 9 10 Very

Why did you eat this snack?

| | | Weight | |
| | | Time | |

D i n n e r

	Amount	Calories	Time:	Location:
Total Dinner Calories:				

How hungry were you before eating?
Not at all 1 2 3 4 5 6 7 8 9 10 **Very**

How do you feel about the healthiness and size of this meal?

large/
unhealthy   small/
                                                    healthy

☹ 🙁 😐 🙂 😄

☐ ☐ ☐ ☐ ☐

Well Done!

S n a c k

	Amount	Calories	Time:	Location:
Total Snack Calories:				

How hungry were you before eating?
Not at all 1 2 3 4 5 6 7 8 9 10 **Very**

Why did you eat this snack?

Total Daily Calories:

E x e r c i s e	Target	Achievement

Are you happy with how you ate and exercised today?

Food	Exercise
☹ 🙁 😐 🙂 😄	☹ 🙁 😐 🙂 😄
☐ ☐ ☐ ☐ ☐	☐ ☐ ☐ ☐ ☐
Well Done!	**Well Done!**

Sunday Week 5

Breakfast	Amount	Calories	Time:	Location:
Total Breakfast Calories:				

How hungry were you before eating?
Not at all 1 2 3 4 5 6 7 8 9 10 Very

How do you feel about the healthiness and size of this meal?

large/ unhealthy   small/ healthy

☐ ☐ ☐ ☐ ☐

Well Done!

Snack	Amount	Calories	Time:	Location:
Total Snack Calories:				

How hungry were you before eating?
Not at all 1 2 3 4 5 6 7 8 9 10 Very

Why did you eat this snack?

Lunch	Amount	Calories	Time:	Location:
Total Lunch Calories:				

How hungry were you before eating?
Not at all 1 2 3 4 5 6 7 8 9 10 Very

How do you feel about the healthiness and size of this meal?

large/ unhealthy   small/ healthy

☐ ☐ ☐ ☐ ☐

Well Done!

Snack	Amount	Calories	Time:	Location:
Total Snack Calories:				

How hungry were you before eating?
Not at all 1 2 3 4 5 6 7 8 9 10 Very

Why did you eat this snack?

Weight				
Time				

D i n n e r

	Amount	Calories	Time:	Location:

How hungry were you before eating?

Not at all 1 2 3 4 5 6 7 8 9 10 Very

How do you feel about the healthiness and size of this meal?

large/unhealthy   small/healthy

☹ ☹ 😐 🙂 😄
☐ ☐ ☐ ☐ ☐

Well Done!

Total Dinner Calories:

S n a c k

	Amount	Calories	Time:	Location:

How hungry were you before eating?

Not at all 1 2 3 4 5 6 7 8 9 10 Very

Why did you eat this snack?

Total Snack Calories:

Total Daily Calories:

E x e r c i s e	Target	Achievement

Are you happy with how you ate and exercised today?

F o o d	E x e r c i s e

☹ ☹ 😐 🙂 😄   ☹ ☹ 😐 🙂 😄
☐ ☐ ☐ ☐ ☐   ☐ ☐ ☐ ☐ ☐

Well Done!   **Well Done!**

Your Weekly Progress

Date: _____

	This week's measurements
Weight:	
Chest:	
Waist:	
Hips:	
Thighs:	
Calves:	
Upper arms:	
Cholesterol:	
Blood pressure:	

How do you feel about this week's progress?

Well Done!

Things you did well this week:

Things you can improve:

Week 6

Monday Week 6

Breakfast

	Amount	Calories	Time:	Location:

How hungry were you before eating?

Not at all 1 2 3 4 5 6 7 8 9 10 Very

How do you feel about the healthiness and size of this meal?

large/unhealthy　　　　　　　　small/healthy

☹ 🙁 😐 🙂 😄

☐ ☐ ☐ ☐ ☐

Total Breakfast Calories:　　　　　　　**Well Done!**

Snack

	Amount	Calories	Time:	Location:

How hungry were you before eating?

Not at all 1 2 3 4 5 6 7 8 9 10 Very

Why did you eat this snack?

Total Snack Calories:

Lunch

	Amount	Calories	Time:	Location:

How hungry were you before eating?

Not at all 1 2 3 4 5 6 7 8 9 10 Very

How do you feel about the healthiness and size of this meal?

large/unhealthy　　　　　　　　small/healthy

☹ 🙁 😐 🙂 😄

☐ ☐ ☐ ☐ ☐

Total Lunch Calories:　　　　　　　**Well Done!**

Snack

	Amount	Calories	Time:	Location:

How hungry were you before eating?

Not at all 1 2 3 4 5 6 7 8 9 10 Very

Why did you eat this snack?

Total Snack Calories:

| | | | Weight | |
| | | | Time | |

Dinner

	Amount	Calories	Time:	Location:
Total Dinner Calories:				

How hungry were you before eating?
Not at all 1 2 3 4 5 6 7 8 9 10 **Very**

How do you feel about the healthiness and size of this meal?

large/unhealthy   small/healthy

☹ ☹ 😐 😀 😄
☐ ☐ ☐ ☐ ☐

Well Done!

Snack

	Amount	Calories	Time:	Location:
Total Snack Calories:				

How hungry were you before eating?
Not at all 1 2 3 4 5 6 7 8 9 10 **Very**

Why did you eat this snack?

Total Daily Calories:

Exercise	Target	Achievement

Are you happy with how you ate and exercised today?

Food

☹ ☹ 😐 😀 😄
☐ ☐ ☐ ☐ ☐

Well Done!

Exercise

☹ ☹ 😐 😀 😄
☐ ☐ ☐ ☐ ☐

Well Done!

Tuesday Week 6

Breakfast

	Amount	Calories	Time:	Location:

Total Breakfast Calories:

How hungry were you before eating?

Not at all   1   2   3   4   5   6   7   8   9   10   Very

How do you feel about the healthiness and size of this meal?

large/unhealthy   small/healthy

☹ 🙁 😐 🙂 😄
☐ ☐ ☐ ☐ ☐

Well Done!

Snack

	Amount	Calories	Time:	Location:

Total Snack Calories:

How hungry were you before eating?

Not at all   1   2   3   4   5   6   7   8   9   10   Very

Why did you eat this snack?

Lunch

	Amount	Calories	Time:	Location:

Total Lunch Calories:

How hungry were you before eating?

Not at all   1   2   3   4   5   6   7   8   9   10   Very

How do you feel about the healthiness and size of this meal?

large/unhealthy   small/healthy

☹ 🙁 😐 🙂 😄
☐ ☐ ☐ ☐ ☐

Well Done!

Snack

	Amount	Calories	Time:	Location:

Total Snack Calories:

How hungry were you before eating?

Not at all   1   2   3   4   5   6   7   8   9   10   Very

Why did you eat this snack?

| | | | | | Weight | |
| Time | | | | | | |

Dinner

	Amount	Calories	Time:	Location:
Total Dinner Calories:				

How hungry were you before eating?
Not at all 1 2 3 4 5 6 7 8 9 10 **Very**

How do you feel about the healthiness and size of this meal?

large/ unhealthy   small/ healthy

☐ ☐ ☐ ☐ ☐

Well Done!

Snack

	Amount	Calories	Time:	Location:
Total Snack Calories:				

How hungry were you before eating?
Not at all 1 2 3 4 5 6 7 8 9 10 **Very**

Why did you eat this snack?

Total Daily Calories:

Exercise	Target	Achievement

Are you happy with how you ate and exercised today?

Food	Exercise
☐ ☐ ☐ ☐ ☐	☐ ☐ ☐ ☐ ☐
Well Done!	**Well Done!**

WednesdayWeek 6

Breakfast	Amount	Calories	Time:	Location:
Total Breakfast Calories:				

How hungry were you before eating?

Not at all   1   2   3   4   5   6   7   8   9   10   Very

How do you feel about the healthiness and size of this meal?

large/unhealthy   small/healthy

☹   ☹   😐   😊   😄

☐   ☐   ☐   ☐   ☐

Well Done!

Snack	Amount	Calories	Time:	Location:
Total Snack Calories:				

How hungry were you before eating?

Not at all   1   2   3   4   5   6   7   8   9   10   Very

Why did you eat this snack?

Lunch	Amount	Calories	Time:	Location:
Total Lunch Calories:				

How hungry were you before eating?

Not at all   1   2   3   4   5   6   7   8   9   10   Very

How do you feel about the healthiness and size of this meal?

large/unhealthy   small/healthy

☹   ☹   😐   😊   😄

☐   ☐   ☐   ☐   ☐

Well Done!

Snack	Amount	Calories	Time:	Location:
Total Snack Calories:				

How hungry were you before eating?

Not at all   1   2   3   4   5   6   7   8   9   10   Very

Why did you eat this snack?

| | | | Weight | |
| | | | Time | |

D i n n e r

	Amount	Calories	Time:	Location:
Total Dinner Calories:				

How hungry were you before eating?

Not at all 1 2 3 4 5 6 7 8 9 10 **Very**

How do you feel about the healthiness and size of this meal?

large/unhealthy   small/healthy

☐ ☐ ☐ ☐ ☐

Well Done!

S n a c k

	Amount	Calories	Time:	Location:
Total Snack Calories:				

How hungry were you before eating?

Not at all 1 2 3 4 5 6 7 8 9 10 **Very**

Why did you eat this snack?

Total Daily Calories:

Exercise	Target	Achievement

Are you happy with how you ate and exercised today?

Food

☐ ☐ ☐ ☐ ☐

Well Done!

Exercise

☐ ☐ ☐ ☐ ☐

Well Done!

Thursday Week 6

Month   Date

Breakfast	Amount	Calories	Time:	Location:

Total Breakfast Calories:

How hungry were you before eating?
Not at all 1 2 3 4 5 6 7 8 9 10 Very

How do you feel about the healthiness and size of this meal?

large/unhealthy   small/healthy

☹ 😞 😐 😊 😄
☐ ☐ ☐ ☐ ☐

Well Done!

Snack	Amount	Calories	Time:	Location:

Total Snack Calories:

How hungry were you before eating?
Not at all 1 2 3 4 5 6 7 8 9 10 Very

Why did you eat this snack?

Lunch	Amount	Calories	Time:	Location:

Total Lunch Calories:

How hungry were you before eating?
Not at all 1 2 3 4 5 6 7 8 9 10 Very

How do you feel about the healthiness and size of this meal?

large/unhealthy   small/healthy

☹ 😞 😐 😊 😄
☐ ☐ ☐ ☐ ☐

Well Done!

Snack	Amount	Calories	Time:	Location:

Total Snack Calories:

How hungry were you before eating?
Not at all 1 2 3 4 5 6 7 8 9 10 Very

Why did you eat this snack?

Weight	
Time	

Dinner

	Amount	Calories	Time:	Location:

Total Dinner Calories:

How hungry were you before eating?

Not at all 1 2 3 4 5 6 7 8 9 10 **Very**

How do you feel about the healthiness and size of this meal?

large/unhealthy   small/healthy

☐   ☐   ☐   ☐   ☐

Well Done!

Snack

	Amount	Calories	Time:	Location:

Total Snack Calories:

How hungry were you before eating?

Not at all 1 2 3 4 5 6 7 8 9 10 **Very**

Why did you eat this snack?

Total Daily Calories:

Exercise	Target	Achievement

Are you happy with how you ate and exercised today?

Food	Exercise
☹ ☹ ☺ ☺ ☺	☹ ☹ ☺ ☺ ☺
☐ ☐ ☐ ☐ ☐	☐ ☐ ☐ ☐ ☐
Well Done!	**Well Done!**

Friday Week 6

Breakfast

	Amount	Calories	Time:	Location:

Total Breakfast Calories:

How hungry were you before eating?

Not at all 1 2 3 4 5 6 7 8 9 10 Very

How do you feel about the healthiness and size of this meal?

large/unhealthy   small/healthy

☐ ☐ ☐ ☐ ☐

Well Done!

Snack

	Amount	Calories	Time:	Location:

Total Snack Calories:

How hungry were you before eating?

Not at all 1 2 3 4 5 6 7 8 9 10 Very

Why did you eat this snack?

Lunch

	Amount	Calories	Time:	Location:

Total Lunch Calories:

How hungry were you before eating?

Not at all 1 2 3 4 5 6 7 8 9 10 Very

How do you feel about the healthiness and size of this meal?

large/unhealthy   small/healthy

☐ ☐ ☐ ☐ ☐

Well Done!

Snack

	Amount	Calories	Time:	Location:

Total Snack Calories:

How hungry were you before eating?

Not at all 1 2 3 4 5 6 7 8 9 10 Very

Why did you eat this snack?

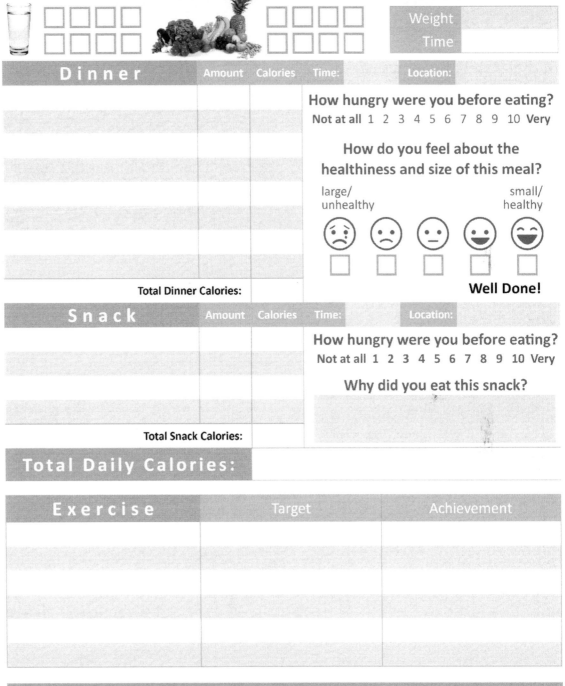

Dinner	Amount	Calories	Time:	Location:
Total Dinner Calories:				

How hungry were you before eating?

Not at all 1 2 3 4 5 6 7 8 9 10 **Very**

How do you feel about the healthiness and size of this meal?

large/unhealthy   small/healthy

☐ ☐ ☐ ☐ ☐

Well Done!

Snack	Amount	Calories	Time:	Location:
Total Snack Calories:				

How hungry were you before eating?

Not at all 1 2 3 4 5 6 7 8 9 10 **Very**

Why did you eat this snack?

Total Daily Calories:

Exercise	Target	Achievement

Are you happy with how you ate and exercised today?

Food

☐ ☐ ☐ ☐ ☐

Well Done!

Exercise

☐ ☐ ☐ ☐ ☐

Well Done!

Breakfast	Amount	Calories	Time:	Location:
Total Breakfast Calories:				

How hungry were you before eating?

Not at all 1 2 3 4 5 6 7 8 9 10 Very

How do you feel about the healthiness and size of this meal?

large/unhealthy   small/healthy

☹ 🙁 😐 🙂 😄
☐ ☐ ☐ ☐ ☐

Well Done!

Snack	Amount	Calories	Time:	Location:
Total Snack Calories:				

How hungry were you before eating?

Not at all 1 2 3 4 5 6 7 8 9 10 Very

Why did you eat this snack?

Lunch	Amount	Calories	Time:	Location:
Total Lunch Calories:				

How hungry were you before eating?

Not at all 1 2 3 4 5 6 7 8 9 10 Very

How do you feel about the healthiness and size of this meal?

large/unhealthy   small/healthy

☹ 🙁 😐 🙂 😄
☐ ☐ ☐ ☐ ☐

Well Done!

Snack	Amount	Calories	Time:	Location:
Total Snack Calories:				

How hungry were you before eating?

Not at all 1 2 3 4 5 6 7 8 9 10 Very

Why did you eat this snack?

Weight	
Time	

Dinner

	Amount	Calories	Time:	Location:
Total Dinner Calories:				

How hungry were you before eating?

Not at all 1 2 3 4 5 6 7 8 9 10 **Very**

How do you feel about the healthiness and size of this meal?

large/unhealthy small/healthy

☹ ☹ 😐 😃 😄
☐ ☐ ☐ ☐ ☐

Well Done!

Snack

	Amount	Calories	Time:	Location:
Total Snack Calories:				

How hungry were you before eating?

Not at all 1 2 3 4 5 6 7 8 9 10 **Very**

Why did you eat this snack?

Total Daily Calories:

Exercise	Target	Achievement

Are you happy with how you ate and exercised today?

Food	Exercise
☹ ☹ 😐 😃 😄	☹ ☹ 😐 😃 😄
☐ ☐ ☐ ☐ ☐	☐ ☐ ☐ ☐ ☐
Well Done!	**Well Done!**

Sunday Week 6

Breakfast	Amount	Calories	Time:	Location:

How hungry were you before eating?

Not at all 1 2 3 4 5 6 7 8 9 10 Very

How do you feel about the healthiness and size of this meal?

large/unhealthy   small/healthy

☹ ☹ 😐 😊 😄
☐ ☐ ☐ ☐ ☐

Well Done!

Total Breakfast Calories:

Snack	Amount	Calories	Time:	Location:

How hungry were you before eating?

Not at all 1 2 3 4 5 6 7 8 9 10 Very

Why did you eat this snack?

Total Snack Calories:

Lunch	Amount	Calories	Time:	Location:

How hungry were you before eating?

Not at all 1 2 3 4 5 6 7 8 9 10 Very

How do you feel about the healthiness and size of this meal?

large/unhealthy   small/healthy

☹ ☹ 😐 😊 😄
☐ ☐ ☐ ☐ ☐

Well Done!

Total Lunch Calories:

Snack	Amount	Calories	Time:	Location:

How hungry were you before eating?

Not at all 1 2 3 4 5 6 7 8 9 10 Very

Why did you eat this snack?

Total Snack Calories:

Weight	
Time	

Dinner

	Amount	Calories	Time:	Location:
Total Dinner Calories:				

How hungry were you before eating?
Not at all 1 2 3 4 5 6 7 8 9 10 Very

How do you feel about the healthiness and size of this meal?

large/
unhealthy

small/
healthy

☹ ☹ 😐 😃 😄
☐ ☐ ☐ ☐ ☐

Well Done!

Snack

	Amount	Calories	Time:	Location:
Total Snack Calories:				

How hungry were you before eating?
Not at all 1 2 3 4 5 6 7 8 9 10 Very

Why did you eat this snack?

Total Daily Calories:

Exercise	Target	Achievement

Are you happy with how you ate and exercised today?

Food	Exercise
☹ ☹ 😐 😃 😄	☹ ☹ 😐 😃 😄
☐ ☐ ☐ ☐ ☐	☐ ☐ ☐ ☐ ☐
Well Done!	Well Done!

Your Weekly Progress

○ ○ ○ ○ ○ ○ ○ ○ ○ ○

Date: _____

	This week's measurements
Weight:	
Chest:	
Waist:	
Hips:	
Thighs:	
Calves:	
Upper arms:	
Cholesterol:	
Blood pressure:	

How do you feel about this week's progress?

☹ ☹ 😐 🙂 😄
☐ ☐ ☐ ☐ ☐
Well Done!

Things you did well this week:

Things you can improve:

Week

7

Monday Week 7

Breakfast	Amount	Calories	Time:	Location:

How hungry were you before eating?

Not at all   1   2   3   4   5   6   7   8   9   10   Very

How do you feel about the healthiness and size of this meal?

large/unhealthy   small/healthy

☐ ☐ ☐ ☐ ☐

Well Done!

Total Breakfast Calories:

Snack	Amount	Calories	Time:	Location:

How hungry were you before eating?

Not at all   1   2   3   4   5   6   7   8   9   10   Very

Why did you eat this snack?

Total Snack Calories:

Lunch	Amount	Calories	Time:	Location:

How hungry were you before eating?

Not at all   1   2   3   4   5   6   7   8   9   10   Very

How do you feel about the healthiness and size of this meal?

large/unhealthy   small/healthy

☐ ☐ ☐ ☐ ☐

Well Done!

Total Lunch Calories:

Snack	Amount	Calories	Time:	Location:

How hungry were you before eating?

Not at all   1   2   3   4   5   6   7   8   9   10   Very

Why did you eat this snack?

Total Snack Calories:

		Weight	
		Time	

Dinner

	Amount	Calories	Time:	Location:
Total Dinner Calories:				

How hungry were you before eating?

Not at all 1 2 3 4 5 6 7 8 9 10 Very

How do you feel about the healthiness and size of this meal?

large/unhealthy   small/healthy

☐ ☐ ☐ ☐ ☐

Well Done!

Snack

	Amount	Calories	Time:	Location:
Total Snack Calories:				

How hungry were you before eating?

Not at all 1 2 3 4 5 6 7 8 9 10 Very

Why did you eat this snack?

Total Daily Calories:

Exercise	Target	Achievement

Are you happy with how you ate and exercised today?

Food	Exercise
☐ ☐ ☐ ☐ ☐	☐ ☐ ☐ ☐ ☐
Well Done!	**Well Done!**

Tuesday Week 7

Breakfast	Amount	Calories	Time:	Location:
Total Breakfast Calories:				

How hungry were you before eating?
Not at all 1 2 3 4 5 6 7 8 9 10 Very

How do you feel about the healthiness and size of this meal?

large/ unhealthy   small/ healthy

☹ ☐   🙁 ☐   😐 ☐   🙂 ☐   😄 ☐

Well Done!

Snack	Amount	Calories	Time:	Location:
Total Snack Calories:				

How hungry were you before eating?
Not at all 1 2 3 4 5 6 7 8 9 10 Very

Why did you eat this snack?

Lunch	Amount	Calories	Time:	Location:
Total Lunch Calories:				

How hungry were you before eating?
Not at all 1 2 3 4 5 6 7 8 9 10 Very

How do you feel about the healthiness and size of this meal?

large/ unhealthy   small/ healthy

☹ ☐   🙁 ☐   😐 ☐   🙂 ☐   😄 ☐

Well Done!

Snack	Amount	Calories	Time:	Location:
Total Snack Calories:				

How hungry were you before eating?
Not at all 1 2 3 4 5 6 7 8 9 10 Very

Why did you eat this snack?

			Weight
			Time

Dinner

	Amount	Calories	Time:	Location:
Total Dinner Calories:				

How hungry were you before eating?
Not at all 1 2 3 4 5 6 7 8 9 10 **Very**

How do you feel about the healthiness and size of this meal?

large/
unhealthy   small/
                                             healthy

☹ ☹ 😐 😃 😄

□ □ □ □ □

Well Done!

Snack

	Amount	Calories	Time:	Location:
Total Snack Calories:				

How hungry were you before eating?
Not at all 1 2 3 4 5 6 7 8 9 10 **Very**

Why did you eat this snack?

Total Daily Calories:

Exercise	Target	Achievement

Are you happy with how you ate and exercised today?

Food	Exercise
☹ ☹ 😐 😃 😄	☹ ☹ 😐 😃 😄
□ □ □ □ □	□ □ □ □ □
Well Done!	**Well Done!**

Wednesday Week 7

Breakfast	Amount	Calories	Time:	Location:
Total Breakfast Calories:				

How hungry were you before eating?
Not at all 1 2 3 4 5 6 7 8 9 10 Very

How do you feel about the healthiness and size of this meal?

large/
unhealthy

small/
healthy

☹ 😦 😐 😃 😄
☐ ☐ ☐ ☐ ☐

Well Done!

Snack	Amount	Calories	Time:	Location:
Total Snack Calories:				

How hungry were you before eating?
Not at all 1 2 3 4 5 6 7 8 9 10 Very

Why did you eat this snack?

Lunch	Amount	Calories	Time:	Location:
Total Lunch Calories:				

How hungry were you before eating?
Not at all 1 2 3 4 5 6 7 8 9 10 Very

How do you feel about the healthiness and size of this meal?

large/
unhealthy

small/
healthy

☹ 😦 😐 😃 😄
☐ ☐ ☐ ☐ ☐

Well Done!

Snack	Amount	Calories	Time:	Location:
Total Snack Calories:				

How hungry were you before eating?
Not at all 1 2 3 4 5 6 7 8 9 10 Very

Why did you eat this snack?

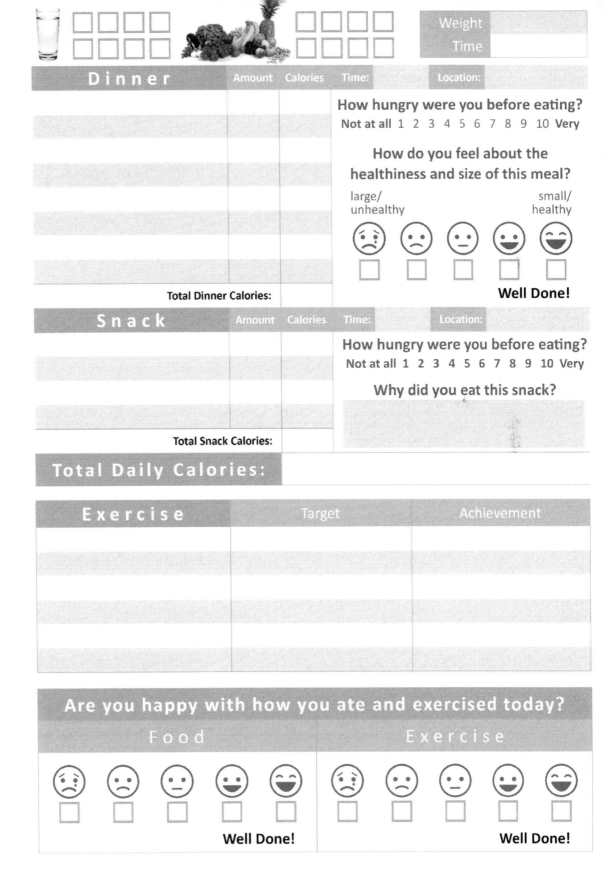

Weight	
Time	

Dinner

	Amount	Calories	Time:	Location:
Total Dinner Calories:				

How hungry were you before eating?
Not at all 1 2 3 4 5 6 7 8 9 10 **Very**

How do you feel about the healthiness and size of this meal?

large/unhealthy   small/healthy

☐ ☐ ☐ ☐ ☐

Well Done!

Snack

	Amount	Calories	Time:	Location:
Total Snack Calories:				

How hungry were you before eating?
Not at all 1 2 3 4 5 6 7 8 9 10 **Very**

Why did you eat this snack?

Total Daily Calories:

Exercise	Target	Achievement

Are you happy with how you ate and exercised today?

Food	Exercise
☐ ☐ ☐ ☐ ☐	☐ ☐ ☐ ☐ ☐
Well Done!	**Well Done!**

Thursday Week 7

Breakfast

	Amount	Calories	Time:	Location:

Total Breakfast Calories:

How hungry were you before eating?

Not at all 1 2 3 4 5 6 7 8 9 10 Very

How do you feel about the healthiness and size of this meal?

large/unhealthy   small/healthy

☐ ☐ ☐ ☐ ☐

Well Done!

Snack

	Amount	Calories	Time:	Location:

Total Snack Calories:

How hungry were you before eating?

Not at all 1 2 3 4 5 6 7 8 9 10 Very

Why did you eat this snack?

Lunch

	Amount	Calories	Time:	Location:

Total Lunch Calories:

How hungry were you before eating?

Not at all 1 2 3 4 5 6 7 8 9 10 Very

How do you feel about the healthiness and size of this meal?

large/unhealthy   small/healthy

☐ ☐ ☐ ☐ ☐

Well Done!

Snack

	Amount	Calories	Time:	Location:

Total Snack Calories:

How hungry were you before eating?

Not at all 1 2 3 4 5 6 7 8 9 10 Very

Why did you eat this snack?

Weight	
Time	

Dinner

	Amount	Calories	Time:	Location:
Total Dinner Calories:				

How hungry were you before eating?

Not at all 1 2 3 4 5 6 7 8 9 10 Very

How do you feel about the healthiness and size of this meal?

large/unhealthy   small/healthy

☹ 😕 😐 😃 😄

☐ ☐ ☐ ☐ ☐

Well Done!

Snack

	Amount	Calories	Time:	Location:
Total Snack Calories:				

How hungry were you before eating?

Not at all 1 2 3 4 5 6 7 8 9 10 Very

Why did you eat this snack?

Total Daily Calories:

Exercise	Target	Achievement

Are you happy with how you ate and exercised today?

Food	Exercise
☹ 😕 😐 😃 😄	☹ 😕 😐 😃 😄
☐ ☐ ☐ ☐ ☐	☐ ☐ ☐ ☐ ☐
Well Done!	**Well Done!**

Friday Week 7

Breakfast

	Amount	Calories	Time:	Location:

How hungry were you before eating?

Not at all 1 2 3 4 5 6 7 8 9 10 Very

How do you feel about the healthiness and size of this meal?

large/unhealthy   small/healthy

☹ ☹ 😐 🙂 😄
☐ ☐ ☐ ☐ ☐

Well Done!

Total Breakfast Calories:

Snack

	Amount	Calories	Time:	Location:

How hungry were you before eating?

Not at all 1 2 3 4 5 6 7 8 9 10 Very

Why did you eat this snack?

Total Snack Calories:

Lunch

	Amount	Calories	Time:	Location:

How hungry were you before eating?

Not at all 1 2 3 4 5 6 7 8 9 10 Very

How do you feel about the healthiness and size of this meal?

large/unhealthy   small/healthy

☹ ☹ 😐 🙂 😄
☐ ☐ ☐ ☐ ☐

Well Done!

Total Lunch Calories:

Snack

	Amount	Calories	Time:	Location:

How hungry were you before eating?

Not at all 1 2 3 4 5 6 7 8 9 10 Very

Why did you eat this snack?

Total Snack Calories:

Weight	
Time	

D i n n e r

	Amount	Calories	Time:	Location:

How hungry were you before eating?

Not at all 1 2 3 4 5 6 7 8 9 10 Very

How do you feel about the healthiness and size of this meal?

large/unhealthy   small/healthy

☐ ☐ ☐ ☐ ☐

Total Dinner Calories:

Well Done!

S n a c k

	Amount	Calories	Time:	Location:

How hungry were you before eating?

Not at all 1 2 3 4 5 6 7 8 9 10 Very

Why did you eat this snack?

Total Snack Calories:

Total Daily Calories:

Exercise	Target	Achievement

Are you happy with how you ate and exercised today?

Food

☐ ☐ ☐ ☐ ☐

Well Done!

Exercise

☐ ☐ ☐ ☐ ☐

Well Done!

Saturday Week 7

Breakfast | Amount | Calories | Time: | Location:

How hungry were you before eating?
Not at all 1 2 3 4 5 6 7 8 9 10 Very

How do you feel about the healthiness and size of this meal?

large/
unhealthy   small/
healthy

☐   ☐   ☐   ☐   ☐

Well Done!

Total Breakfast Calories:

Snack | Amount | Calories | Time: | Location:

How hungry were you before eating?
Not at all 1 2 3 4 5 6 7 8 9 10 Very

Why did you eat this snack?

Total Snack Calories:

Lunch | Amount | Calories | Time: | Location:

How hungry were you before eating?
Not at all 1 2 3 4 5 6 7 8 9 10 Very

How do you feel about the healthiness and size of this meal?

large/
unhealthy   small/
healthy

☐   ☐   ☐   ☐   ☐

Well Done!

Total Lunch Calories:

Snack | Amount | Calories | Time: | Location:

How hungry were you before eating?
Not at all 1 2 3 4 5 6 7 8 9 10 Very

Why did you eat this snack?

Total Snack Calories:

D i n n e r

	Amount	Calories	Time:	Location:
Total Dinner Calories:				

How hungry were you before eating?

Not at all 1 2 3 4 5 6 7 8 9 10 Very

How do you feel about the healthiness and size of this meal?

large/unhealthy   small/healthy

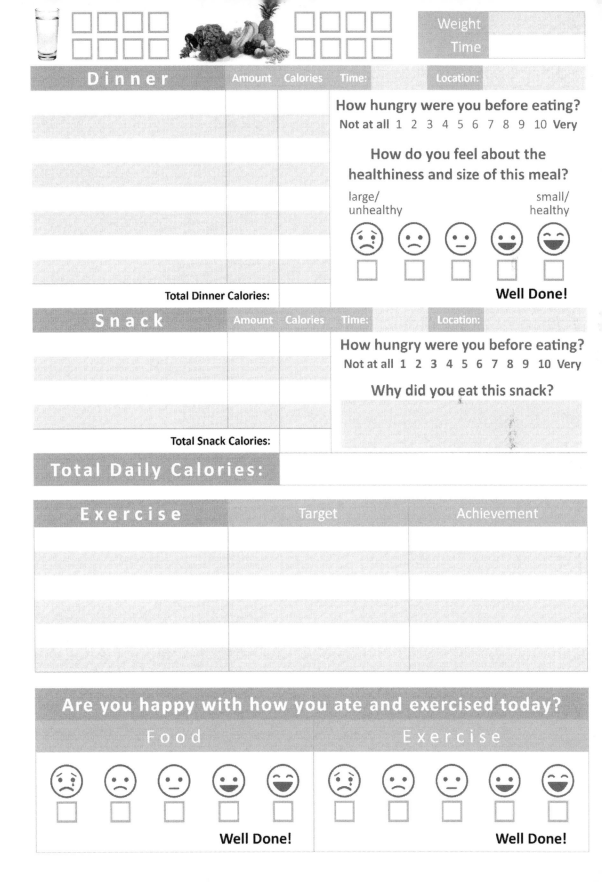

Well Done!

S n a c k

	Amount	Calories	Time:	Location:
Total Snack Calories:				

How hungry were you before eating?

Not at all 1 2 3 4 5 6 7 8 9 10 Very

Why did you eat this snack?

Total Daily Calories:

E x e r c i s e	Target	Achievement

Are you happy with how you ate and exercised today?

F o o d

Well Done!

E x e r c i s e

Well Done!

Sunday Week 7

Breakfast	Amount	Calories	Time:	Location:

Total Breakfast Calories:

How hungry were you before eating?
Not at all 1 2 3 4 5 6 7 8 9 10 Very

How do you feel about the healthiness and size of this meal?

large/ unhealthy   small/ healthy

☐ ☐ ☐ ☐ ☐

Well Done!

Snack	Amount	Calories	Time:	Location:

Total Snack Calories:

How hungry were you before eating?
Not at all 1 2 3 4 5 6 7 8 9 10 Very

Why did you eat this snack?

Lunch	Amount	Calories	Time:	Location:

Total Lunch Calories:

How hungry were you before eating?
Not at all 1 2 3 4 5 6 7 8 9 10 Very

How do you feel about the healthiness and size of this meal?

large/ unhealthy   small/ healthy

☐ ☐ ☐ ☐ ☐

Well Done!

Snack	Amount	Calories	Time:	Location:

Total Snack Calories:

How hungry were you before eating?
Not at all 1 2 3 4 5 6 7 8 9 10 Very

Why did you eat this snack?

	Weight	
	Time	

Dinner

	Amount	Calories	Time:	Location:
Total Dinner Calories:				

How hungry were you before eating?
Not at all 1 2 3 4 5 6 7 8 9 10 **Very**

How do you feel about the healthiness and size of this meal?

large/unhealthy   small/healthy

☹ 🙁 😐 🙂 😄

☐ ☐ ☐ ☐ ☐

Well Done!

Snack

	Amount	Calories	Time:	Location:
Total Snack Calories:				

How hungry were you before eating?
Not at all 1 2 3 4 5 6 7 8 9 10 **Very**

Why did you eat this snack?

Total Daily Calories:

Exercise	Target	Achievement

Are you happy with how you ate and exercised today?

Food

☹ 🙁 😐 🙂 😄

☐ ☐ ☐ ☐ ☐

Well Done!

Exercise

☹ 🙁 😐 🙂 😄

☐ ☐ ☐ ☐ ☐

Well Done!

Your Weekly Progress

○ ○ ○ ○ ○ ○ ○ ○ ○ ○ ○

Date: _____

	This week's measurements
Weight:	
Chest:	
Waist:	
Hips:	
Thighs:	
Calves:	
Upper arms:	
Cholesterol:	
Blood pressure:	

How do you feel about this week's progress?

☐ ☐ ☐ ☐ ☐

Well Done!

Things you did well this week:

Things you can improve:

Week 8

Monday Week 8

Breakfast | Amount | Calories | Time: | Location:

	Amount	Calories	Time:	Location:
Total Breakfast Calories:				

How hungry were you before eating?
Not at all 1 2 3 4 5 6 7 8 9 10 Very

How do you feel about the healthiness and size of this meal?

large/unhealthy   small/healthy

☐ ☐ ☐ ☐ ☐

Well Done!

Snack | Amount | Calories | Time: | Location:

	Amount	Calories	Time:	Location:
Total Snack Calories:				

How hungry were you before eating?
Not at all 1 2 3 4 5 6 7 8 9 10 Very

Why did you eat this snack?

Lunch | Amount | Calories | Time: | Location:

	Amount	Calories	Time:	Location:
Total Lunch Calories:				

How hungry were you before eating?
Not at all 1 2 3 4 5 6 7 8 9 10 Very

How do you feel about the healthiness and size of this meal?

large/unhealthy   small/healthy

☐ ☐ ☐ ☐ ☐

Well Done!

Snack | Amount | Calories | Time: | Location:

	Amount	Calories	Time:	Location:
Total Snack Calories:				

How hungry were you before eating?
Not at all 1 2 3 4 5 6 7 8 9 10 Very

Why did you eat this snack?

						Weight	
						Time	

D i n n e r

	Amount	Calories	Time:	Location:

How hungry were you before eating?

Not at all 1 2 3 4 5 6 7 8 9 10 **Very**

How do you feel about the healthiness and size of this meal?

large/unhealthy   small/healthy

☹ ☹ 😐 🙂 😄
☐ ☐ ☐ ☐ ☐

Total Dinner Calories:

Well Done!

S n a c k

	Amount	Calories	Time:	Location:

How hungry were you before eating?

Not at all 1 2 3 4 5 6 7 8 9 10 **Very**

Why did you eat this snack?

Total Snack Calories:

Total Daily Calories:

E x e r c i s e	Target	Achievement

Are you happy with how you ate and exercised today?

F o o d	E x e r c i s e

☹ ☹ 😐 🙂 😄   ☹ ☹ 😐 🙂 😄
☐ ☐ ☐ ☐ ☐   ☐ ☐ ☐ ☐ ☐

Well Done!   **Well Done!**

Tuesday Week 8

Breakfast　　Amount　Calories　Time:　　Location:

How hungry were you before eating?

Not at all 1 2 3 4 5 6 7 8 9 10 Very

How do you feel about the
healthiness and size of this meal?

large/
unhealthy

small/
healthy

☹ ☹ 😐 🙂 😄
☐ ☐ ☐ ☐ ☐

Well Done!

Total Breakfast Calories:

Snack　　Amount　Calories　Time:　　Location:

How hungry were you before eating?

Not at all 1 2 3 4 5 6 7 8 9 10 Very

Why did you eat this snack?

Total Snack Calories:

Lunch　　Amount　Calories　Time:　　Location:

How hungry were you before eating?

Not at all 1 2 3 4 5 6 7 8 9 10 Very

How do you feel about the
healthiness and size of this meal?

large/
unhealthy

small/
healthy

☹ ☹ 😐 🙂 😄
☐ ☐ ☐ ☐ ☐

Well Done!

Total Lunch Calories:

Snack　　Amount　Calories　Time:　　Location:

How hungry were you before eating?

Not at all 1 2 3 4 5 6 7 8 9 10 Very

Why did you eat this snack?

Total Snack Calories:

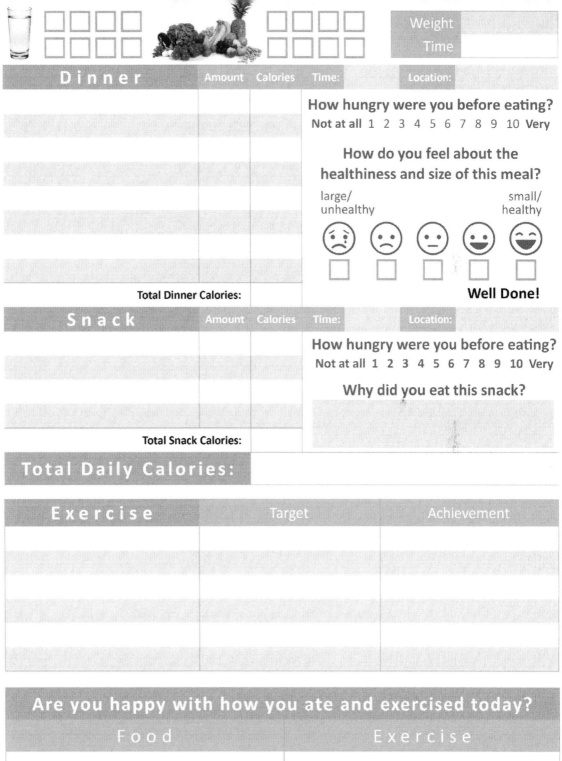

□□□□		□□□□	Weight
□□□□		□□□□	Time

D i n n e r

	Amount	Calories	Time:	Location:

How hungry were you before eating?
Not at all 1 2 3 4 5 6 7 8 9 10 Very

How do you feel about the healthiness and size of this meal?

large/unhealthy   small/healthy

□   □   □   □   □

Well Done!

Total Dinner Calories:

S n a c k

	Amount	Calories	Time:	Location:

How hungry were you before eating?
Not at all 1 2 3 4 5 6 7 8 9 10 Very

Why did you eat this snack?

Total Snack Calories:

Total Daily Calories:

Exercise	Target	Achievement

Are you happy with how you ate and exercised today?

Food

□   □   □   □   □

Well Done!

Exercise

□   □   □   □   □

Well Done!

Wednesday Week 8

Month   Date

Breakfast

	Amount	Calories	Time:	Location:
Total Breakfast Calories:				

How hungry were you before eating?
Not at all 1 2 3 4 5 6 7 8 9 10 Very

How do you feel about the healthiness and size of this meal?

large/unhealthy   small/healthy

☐ ☐ ☐ ☐ ☐

Well Done!

Snack

	Amount	Calories	Time:	Location:
Total Snack Calories:				

How hungry were you before eating?
Not at all 1 2 3 4 5 6 7 8 9 10 Very

Why did you eat this snack?

Lunch

	Amount	Calories	Time:	Location:
Total Lunch Calories:				

How hungry were you before eating?
Not at all 1 2 3 4 5 6 7 8 9 10 Very

How do you feel about the healthiness and size of this meal?

large/unhealthy   small/healthy

☐ ☐ ☐ ☐ ☐

Well Done!

Snack

	Amount	Calories	Time:	Location:
Total Snack Calories:				

How hungry were you before eating?
Not at all 1 2 3 4 5 6 7 8 9 10 Very

Why did you eat this snack?

		Weight
		Time

Dinner	Amount	Calories	Time:	Location:
Total Dinner Calories:				

How hungry were you before eating?
Not at all 1 2 3 4 5 6 7 8 9 10 Very

How do you feel about the healthiness and size of this meal?

large/unhealthy   small/healthy

☹ 🙁 😐 🙂 😄
☐ ☐ ☐ ☐ ☐

Well Done!

Snack	Amount	Calories	Time:	Location:
Total Snack Calories:				

How hungry were you before eating?
Not at all 1 2 3 4 5 6 7 8 9 10 Very

Why did you eat this snack?

Total Daily Calories:

Exercise	Target	Achievement

Are you happy with how you ate and exercised today?

Food	Exercise
☹ 🙁 😐 🙂 😄	☹ 🙁 😐 🙂 😄
☐ ☐ ☐ ☐ ☐	☐ ☐ ☐ ☐ ☐
Well Done!	**Well Done!**

Thursday Week 8

Breakfast

	Amount	Calories	Time:	Location:
Total Breakfast Calories:				

How hungry were you before eating?

Not at all 1 2 3 4 5 6 7 8 9 10 Very

How do you feel about the healthiness and size of this meal?

large/unhealthy   small/healthy

☹ 😦 😐 🙂 😄

☐ ☐ ☐ ☐ ☐

Well Done!

Snack

	Amount	Calories	Time:	Location:
Total Snack Calories:				

How hungry were you before eating?

Not at all 1 2 3 4 5 6 7 8 9 10 Very

Why did you eat this snack?

Lunch

	Amount	Calories	Time:	Location:
Total Lunch Calories:				

How hungry were you before eating?

Not at all 1 2 3 4 5 6 7 8 9 10 Very

How do you feel about the healthiness and size of this meal?

large/unhealthy   small/healthy

☹ 😦 😐 🙂 😄

☐ ☐ ☐ ☐ ☐

Well Done!

Snack

	Amount	Calories	Time:	Location:
Total Snack Calories:				

How hungry were you before eating?

Not at all 1 2 3 4 5 6 7 8 9 10 Very

Why did you eat this snack?

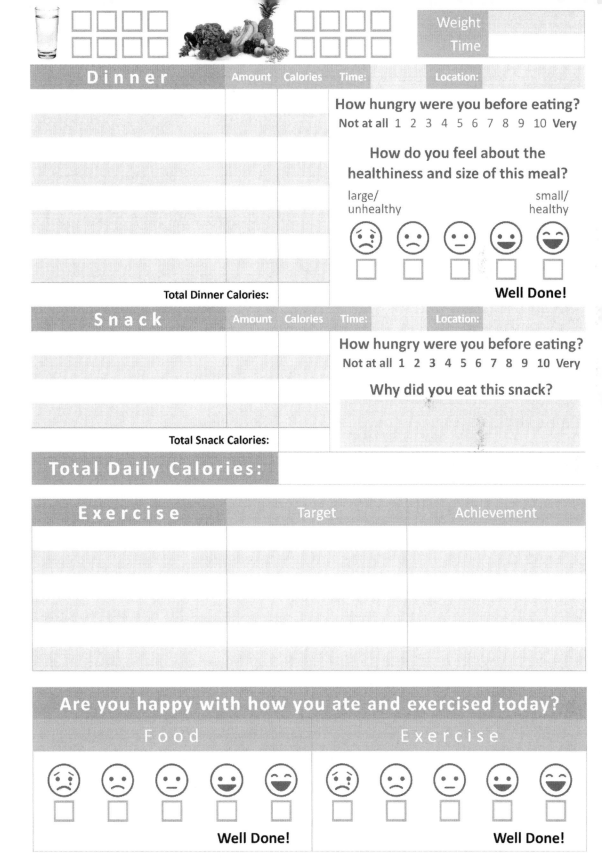

	Weight	
	Time	

Dinner

	Amount	Calories	Time:	Location:
Total Dinner Calories:				

How hungry were you before eating?
Not at all 1 2 3 4 5 6 7 8 9 10 Very

How do you feel about the healthiness and size of this meal?

large/unhealthy — small/healthy

☐ ☐ ☐ ☐ ☐

Well Done!

Snack

	Amount	Calories	Time:	Location:
Total Snack Calories:				

How hungry were you before eating?
Not at all 1 2 3 4 5 6 7 8 9 10 Very

Why did you eat this snack?

Total Daily Calories:

Exercise	Target	Achievement

Are you happy with how you ate and exercised today?

Food	Exercise
☐ ☐ ☐ ☐ ☐	☐ ☐ ☐ ☐ ☐
Well Done!	**Well Done!**

FridayWeek 8

Breakfast	Amount	Calories	Time:	Location:

Total Breakfast Calories:

How hungry were you before eating?

Not at all 1 2 3 4 5 6 7 8 9 10 Very

How do you feel about the healthiness and size of this meal?

large/ unhealthy   small/ healthy

☹ ☹ 😐 😊 😄
☐ ☐ ☐ ☐ ☐

Well Done!

Snack	Amount	Calories	Time:	Location:

Total Snack Calories:

How hungry were you before eating?

Not at all 1 2 3 4 5 6 7 8 9 10 Very

Why did you eat this snack?

Lunch	Amount	Calories	Time:	Location:

Total Lunch Calories:

How hungry were you before eating?

Not at all 1 2 3 4 5 6 7 8 9 10 Very

How do you feel about the healthiness and size of this meal?

large/ unhealthy   small/ healthy

☹ ☹ 😐 😊 😄
☐ ☐ ☐ ☐ ☐

Well Done!

Snack	Amount	Calories	Time:	Location:

Total Snack Calories:

How hungry were you before eating?

Not at all 1 2 3 4 5 6 7 8 9 10 Very

Why did you eat this snack?

					Weight	
					Time	

Dinner

	Amount	Calories	Time:	Location:
Total Dinner Calories:				

How hungry were you before eating?
Not at all 1 2 3 4 5 6 7 8 9 10 **Very**

How do you feel about the healthiness and size of this meal?

large/unhealthy   small/healthy

☹ ☹ 😐 🙂 😄
☐ ☐ ☐ ☐ ☐

Well Done!

Snack

	Amount	Calories	Time:	Location:
Total Snack Calories:				

How hungry were you before eating?
Not at all 1 2 3 4 5 6 7 8 9 10 **Very**

Why did you eat this snack?

Total Daily Calories:

Exercise	Target	Achievement

Are you happy with how you ate and exercised today?

Food

☹ ☹ 😐 🙂 😄
☐ ☐ ☐ ☐ ☐

Well Done!

Exercise

☹ ☹ 😐 🙂 😄
☐ ☐ ☐ ☐ ☐

Well Done!

Saturday Week 8

Breakfast | Amount | Calories | Time: | Location:

How hungry were you before eating?

Not at all 1 2 3 4 5 6 7 8 9 10 Very

How do you feel about the healthiness and size of this meal?

large/
unhealthy

small/
healthy

☐ ☐ ☐ ☐ ☐

Well Done!

Total Breakfast Calories:

Snack | Amount | Calories | Time: | Location:

How hungry were you before eating?

Not at all 1 2 3 4 5 6 7 8 9 10 Very

Why did you eat this snack?

Total Snack Calories:

Lunch | Amount | Calories | Time: | Location:

How hungry were you before eating?

Not at all 1 2 3 4 5 6 7 8 9 10 Very

How do you feel about the healthiness and size of this meal?

large/
unhealthy

small/
healthy

☐ ☐ ☐ ☐ ☐

Well Done!

Total Lunch Calories:

Snack | Amount | Calories | Time: | Location:

How hungry were you before eating?

Not at all 1 2 3 4 5 6 7 8 9 10 Very

Why did you eat this snack?

Total Snack Calories:

					Weight	
					Time	

Dinner

	Amount	Calories	Time:	Location:

How hungry were you before eating?

Not at all 1 2 3 4 5 6 7 8 9 10 Very

How do you feel about the healthiness and size of this meal?

large/ unhealthy   small/ healthy

☐ ☐ ☐ ☐ ☐

Total Dinner Calories:

Well Done!

Snack

	Amount	Calories	Time:	Location:

How hungry were you before eating?

Not at all 1 2 3 4 5 6 7 8 9 10 Very

Why did you eat this snack?

Total Snack Calories:

Total Daily Calories:

Exercise	Target	Achievement

Are you happy with how you ate and exercised today?

Food	Exercise
☐ ☐ ☐ ☐ ☐	☐ ☐ ☐ ☐ ☐
Well Done!	Well Done!

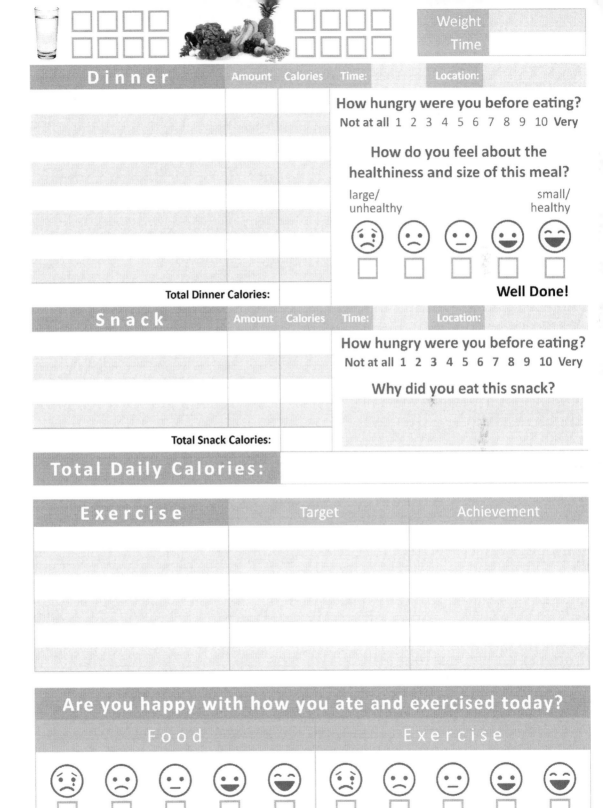

Sunday Week 8

Breakfast

	Amount	Calories	Time:	Location:

Total Breakfast Calories:

How hungry were you before eating?

Not at all 1 2 3 4 5 6 7 8 9 10 Very

How do you feel about the healthiness and size of this meal?

large/unhealthy   small/healthy

☐ ☐ ☐ ☐ ☐

Well Done!

Snack

	Amount	Calories	Time:	Location:

Total Snack Calories:

How hungry were you before eating?

Not at all 1 2 3 4 5 6 7 8 9 10 Very

Why did you eat this snack?

Lunch

	Amount	Calories	Time:	Location:

Total Lunch Calories:

How hungry were you before eating?

Not at all 1 2 3 4 5 6 7 8 9 10 Very

How do you feel about the healthiness and size of this meal?

large/unhealthy   small/healthy

☐ ☐ ☐ ☐ ☐

Well Done!

Snack

	Amount	Calories	Time:	Location:

Total Snack Calories:

How hungry were you before eating?

Not at all 1 2 3 4 5 6 7 8 9 10 Very

Why did you eat this snack?

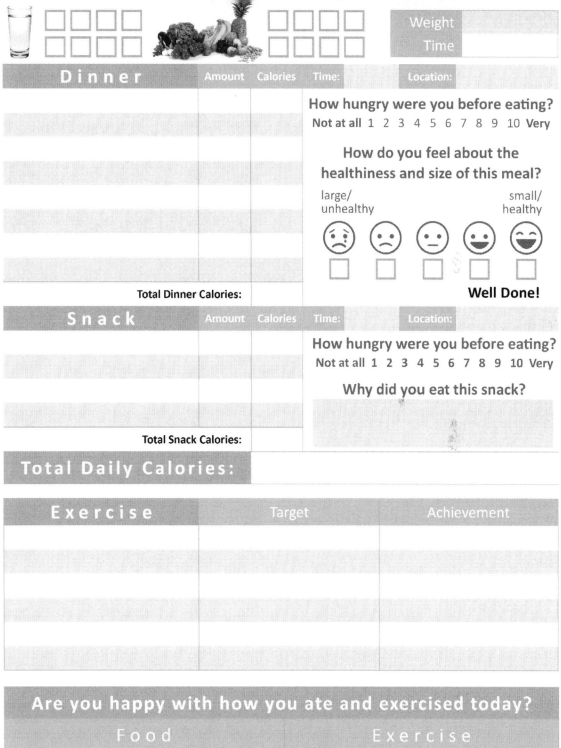

D i n n e r	Amount	Calories	Time:	Location:
Total Dinner Calories:				

How hungry were you before eating?

Not at all 1 2 3 4 5 6 7 8 9 10 Very

How do you feel about the healthiness and size of this meal?

large/unhealthy　　　　　small/healthy

☐ ☐ ☐ ☐ ☐

Well Done!

S n a c k	Amount	Calories	Time:	Location:
Total Snack Calories:				

How hungry were you before eating?

Not at all 1 2 3 4 5 6 7 8 9 10 Very

Why did you eat this snack?

Total Daily Calories:

E x e r c i s e	Target	Achievement

Are you happy with how you ate and exercised today?

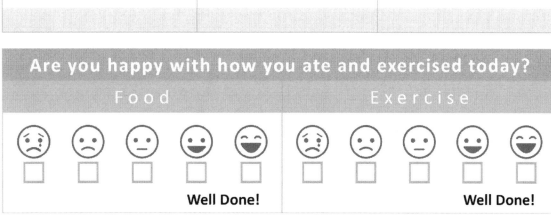

Food	Exercise
☐ ☐ ☐ ☐ ☐	☐ ☐ ☐ ☐ ☐
Well Done!	Well Done!

Your Weekly Progress

○○○○○○○○○○○

Date: _____

	This week's measurements
Weight:	
Chest:	
Waist:	
Hips:	
Thighs:	
Calves:	
Upper arms:	
Cholesterol:	
Blood pressure:	

How do you feel about this week's progress?

Well Done!

Things you did well this week:

Things you can improve:

Week 9

Monday Week 9

Breakfast	Amount	Calories	Time:	Location:
Total Breakfast Calories:				

How hungry were you before eating?
Not at all 1 2 3 4 5 6 7 8 9 10 Very

How do you feel about the healthiness and size of this meal?

large/unhealthy   small/healthy

☐   ☐   ☐   ☐   ☐

Well Done!

Snack	Amount	Calories	Time:	Location:
Total Snack Calories:				

How hungry were you before eating?
Not at all 1 2 3 4 5 6 7 8 9 10 Very

Why did you eat this snack?

Lunch	Amount	Calories	Time:	Location:
Total Lunch Calories:				

How hungry were you before eating?
Not at all 1 2 3 4 5 6 7 8 9 10 Very

How do you feel about the healthiness and size of this meal?

large/unhealthy   small/healthy

☐   ☐   ☐   ☐   ☐

Well Done!

Snack	Amount	Calories	Time:	Location:
Total Snack Calories:				

How hungry were you before eating?
Not at all 1 2 3 4 5 6 7 8 9 10 Very

Why did you eat this snack?

| | | | | Weight | |
| | | | | Time | |

Dinner

	Amount	Calories	Time:	Location:
Total Dinner Calories:				

How hungry were you before eating?

Not at all 1 2 3 4 5 6 7 8 9 10 Very

How do you feel about the healthiness and size of this meal?

large/unhealthy   small/healthy

☹ ☹ 😐 🙂 😄

☐ ☐ ☐ ☐ ☐

Well Done!

Snack

	Amount	Calories	Time:	Location:
Total Snack Calories:				

How hungry were you before eating?

Not at all 1 2 3 4 5 6 7 8 9 10 Very

Why did you eat this snack?

Total Daily Calories:

Exercise	Target	Achievement

Are you happy with how you ate and exercised today?

Food	Exercise
☹ ☹ 😐 🙂 😄	☹ ☹ 😐 🙂 😄
☐ ☐ ☐ ☐ ☐	☐ ☐ ☐ ☐ ☐
Well Done!	**Well Done!**

Tuesday Week 9

Breakfast	Amount	Calories	Time:	Location:
Total Breakfast Calories:				

How hungry were you before eating?
Not at all 1 2 3 4 5 6 7 8 9 10 Very

How do you feel about the healthiness and size of this meal?

large/unhealthy   small/healthy

☹ ☐   🙁 ☐   😐 ☐   🙂 ☐   😄 ☐

Well Done!

Snack	Amount	Calories	Time:	Location:
Total Snack Calories:				

How hungry were you before eating?
Not at all 1 2 3 4 5 6 7 8 9 10 Very

Why did you eat this snack?

Lunch	Amount	Calories	Time:	Location:
Total Lunch Calories:				

How hungry were you before eating?
Not at all 1 2 3 4 5 6 7 8 9 10 Very

How do you feel about the healthiness and size of this meal?

large/unhealthy   small/healthy

☹ ☐   🙁 ☐   😐 ☐   🙂 ☐   😄 ☐

Well Done!

Snack	Amount	Calories	Time:	Location:
Total Snack Calories:				

How hungry were you before eating?
Not at all 1 2 3 4 5 6 7 8 9 10 Very

Why did you eat this snack?

				Weight	
				Time	

Dinner

	Amount	Calories	Time:	Location:

How hungry were you before eating?

Not at all 1 2 3 4 5 6 7 8 9 10 **Very**

How do you feel about the healthiness and size of this meal?

large/unhealthy   small/healthy

☹ 😕 😐 🙂 😄

☐ ☐ ☐ ☐ ☐

Well Done!

Total Dinner Calories:

Snack

	Amount	Calories	Time:	Location:

How hungry were you before eating?

Not at all 1 2 3 4 5 6 7 8 9 10 **Very**

Why did you eat this snack?

Total Snack Calories:

Total Daily Calories:

Exercise	Target	Achievement

Are you happy with how you ate and exercised today?

Food	Exercise
☹ 😕 😐 🙂 😄	☹ 😕 😐 🙂 😄
☐ ☐ ☐ ☐ ☐	☐ ☐ ☐ ☐ ☐
Well Done!	**Well Done!**

Wednesday Week 9

Breakfast
	Amount	Calories	Time:	Location:

Total Breakfast Calories:

How hungry were you before eating?
Not at all 1 2 3 4 5 6 7 8 9 10 Very

How do you feel about the healthiness and size of this meal?
large/unhealthy   small/healthy

☐ ☐ ☐ ☐ ☐

Well Done!

Snack
	Amount	Calories	Time:	Location:

Total Snack Calories:

How hungry were you before eating?
Not at all 1 2 3 4 5 6 7 8 9 10 Very

Why did you eat this snack?

Lunch
	Amount	Calories	Time:	Location:

Total Lunch Calories:

How hungry were you before eating?
Not at all 1 2 3 4 5 6 7 8 9 10 Very

How do you feel about the healthiness and size of this meal?
large/unhealthy   small/healthy

☐ ☐ ☐ ☐ ☐

Well Done!

Snack
	Amount	Calories	Time:	Location:

Total Snack Calories:

How hungry were you before eating?
Not at all 1 2 3 4 5 6 7 8 9 10 Very

Why did you eat this snack?

			Weight	
			Time	

Dinner

	Amount	Calories	Time:	Location:
Total Dinner Calories:				

How hungry were you before eating?
Not at all 1 2 3 4 5 6 7 8 9 10 Very

How do you feel about the healthiness and size of this meal?

large/unhealthy   small/healthy

☐   ☐   ☐   ☐   ☐

Well Done!

Snack

	Amount	Calories	Time:	Location:
Total Snack Calories:				

How hungry were you before eating?
Not at all 1 2 3 4 5 6 7 8 9 10 Very

Why did you eat this snack?

Total Daily Calories:

Exercise	Target	Achievement

Are you happy with how you ate and exercised today?

Food

☐   ☐   ☐   ☐   ☐

Well Done!

Exercise

☐   ☐   ☐   ☐   ☐

Well Done!

Thursday Week 9

Breakfast

	Amount	Calories	Time:	Location:

Total Breakfast Calories:

How hungry were you before eating?

Not at all 1 2 3 4 5 6 7 8 9 10 Very

How do you feel about the healthiness and size of this meal?

large/unhealthy   small/healthy

☐ ☐ ☐ ☐ ☐

Well Done!

Snack

	Amount	Calories	Time:	Location:

Total Snack Calories:

How hungry were you before eating?

Not at all 1 2 3 4 5 6 7 8 9 10 Very

Why did you eat this snack?

Lunch

	Amount	Calories	Time:	Location:

Total Lunch Calories:

How hungry were you before eating?

Not at all 1 2 3 4 5 6 7 8 9 10 Very

How do you feel about the healthiness and size of this meal?

large/unhealthy   small/healthy

☐ ☐ ☐ ☐ ☐

Well Done!

Snack

	Amount	Calories	Time:	Location:

Total Snack Calories:

How hungry were you before eating?

Not at all 1 2 3 4 5 6 7 8 9 10 Very

Why did you eat this snack?

	Weight	
	Time	

Dinner

	Amount	Calories	Time:	Location:
Total Dinner Calories:				

How hungry were you before eating?
Not at all 1 2 3 4 5 6 7 8 9 10 Very

How do you feel about the healthiness and size of this meal?

large/unhealthy   small/healthy

☐ ☐ ☐ ☐ ☐

Well Done!

Snack

	Amount	Calories	Time:	Location:
Total Snack Calories:				

How hungry were you before eating?
Not at all 1 2 3 4 5 6 7 8 9 10 Very

Why did you eat this snack?

Total Daily Calories:

Exercise	Target	Achievement

Are you happy with how you ate and exercised today?

Food	Exercise
☐ ☐ ☐ ☐ ☐	☐ ☐ ☐ ☐ ☐
Well Done!	**Well Done!**

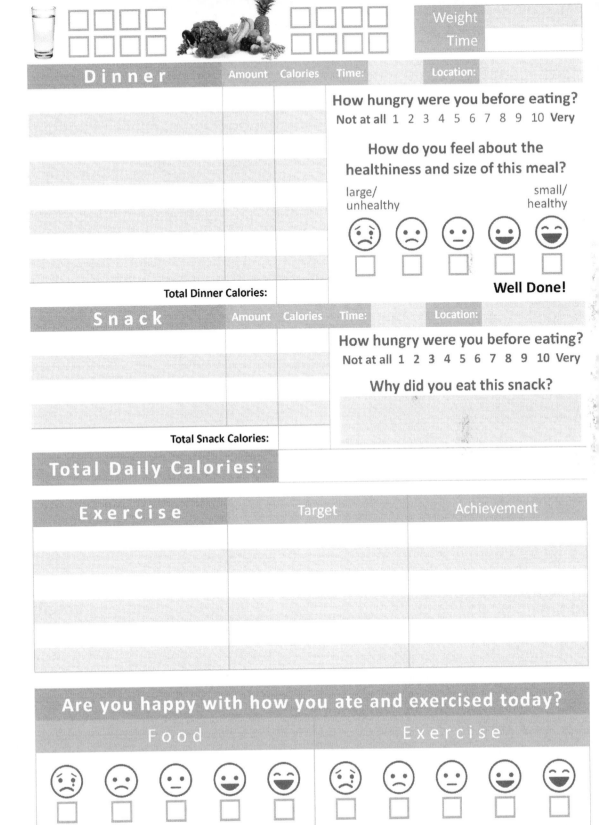

Friday Week 9

Month   Date

Breakfast

	Amount	Calories	Time:	Location:
Total Breakfast Calories:				

How hungry were you before eating?

Not at all 1 2 3 4 5 6 7 8 9 10 Very

How do you feel about the healthiness and size of this meal?

large/unhealthy   small/healthy

☐ ☐ ☐ ☐ ☐

Well Done!

Snack

	Amount	Calories	Time:	Location:
Total Snack Calories:				

How hungry were you before eating?

Not at all 1 2 3 4 5 6 7 8 9 10 Very

Why did you eat this snack?

Lunch

	Amount	Calories	Time:	Location:
Total Lunch Calories:				

How hungry were you before eating?

Not at all 1 2 3 4 5 6 7 8 9 10 Very

How do you feel about the healthiness and size of this meal?

large/unhealthy   small/healthy

☐ ☐ ☐ ☐ ☐

Well Done!

Snack

	Amount	Calories	Time:	Location:
Total Snack Calories:				

How hungry were you before eating?

Not at all 1 2 3 4 5 6 7 8 9 10 Very

Why did you eat this snack?

						Weight	
						Time	

D i n n e r		Amount	Calories	Time:		Location:	

How hungry were you before eating?
Not at all 1 2 3 4 5 6 7 8 9 10 Very

How do you feel about the healthiness and size of this meal?

large/
unhealthy

small/
healthy

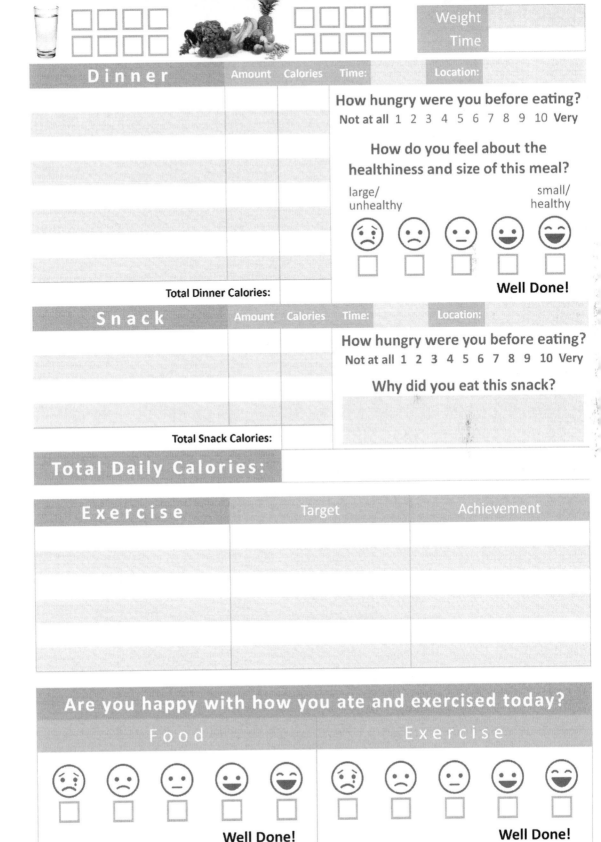

☐ ☐ ☐ ☐ ☐

Well Done!

Total Dinner Calories:

S n a c k	Amount	Calories	Time:	Location:

How hungry were you before eating?
Not at all 1 2 3 4 5 6 7 8 9 10 Very

Why did you eat this snack?

Total Snack Calories:

Total Daily Calories:

E x e r c i s e	Target	Achievement

Are you happy with how you ate and exercised today?

F o o d	**E x e r c i s e**
☐ ☐ ☐ ☐ ☐	☐ ☐ ☐ ☐ ☐
Well Done!	**Well Done!**

Saturday Week 9

Breakfast

	Amount	Calories	Time:	Location:

How hungry were you before eating?

Not at all 1 2 3 4 5 6 7 8 9 10 Very

How do you feel about the healthiness and size of this meal?

large/unhealthy   small/healthy

☐ ☐ ☐ ☐ ☐

Well Done!

Total Breakfast Calories:

Snack

	Amount	Calories	Time:	Location:

How hungry were you before eating?

Not at all 1 2 3 4 5 6 7 8 9 10 Very

Why did you eat this snack?

Total Snack Calories:

Lunch

	Amount	Calories	Time:	Location:

How hungry were you before eating?

Not at all 1 2 3 4 5 6 7 8 9 10 Very

How do you feel about the healthiness and size of this meal?

large/unhealthy   small/healthy

☐ ☐ ☐ ☐ ☐

Well Done!

Total Lunch Calories:

Snack

	Amount	Calories	Time:	Location:

How hungry were you before eating?

Not at all 1 2 3 4 5 6 7 8 9 10 Very

Why did you eat this snack?

Total Snack Calories:

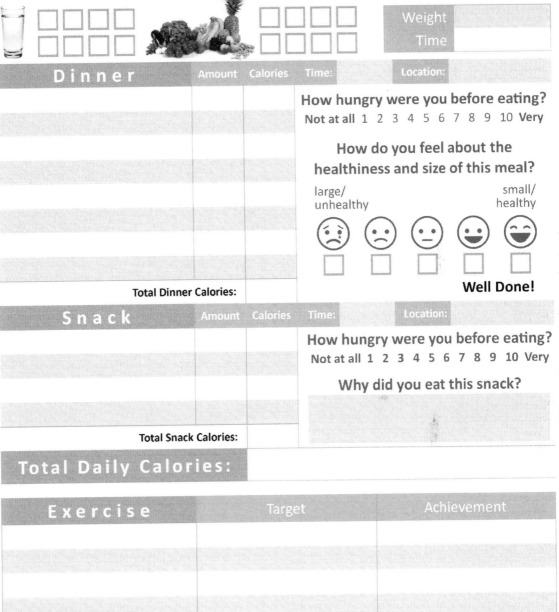

D i n n e r

	Amount	Calories	Time:	Location:
Total Dinner Calories:				

How hungry were you before eating?

Not at all 1 2 3 4 5 6 7 8 9 10 **Very**

How do you feel about the healthiness and size of this meal?

large/
unhealthy

small/
healthy

Well Done!

S n a c k

	Amount	Calories	Time:	Location:
Total Snack Calories:				

How hungry were you before eating?

Not at all 1 2 3 4 5 6 7 8 9 10 **Very**

Why did you eat this snack?

Total Daily Calories:

Exercise	Target	Achievement

Are you happy with how you ate and exercised today?

Food

Exercise

Well Done!

Well Done!

Sunday Week 9

Breakfast | Amount | Calories | Time: | Location:

How hungry were you before eating?

Not at all 1 2 3 4 5 6 7 8 9 10 Very

How do you feel about the healthiness and size of this meal?

large/ unhealthy   small/ healthy

☹ ☹ 😐 😊 😄
☐ ☐ ☐ ☐ ☐

Well Done!

Total Breakfast Calories:

Snack | Amount | Calories | Time: | Location:

How hungry were you before eating?

Not at all 1 2 3 4 5 6 7 8 9 10 Very

Why did you eat this snack?

Total Snack Calories:

Lunch | Amount | Calories | Time: | Location:

How hungry were you before eating?

Not at all 1 2 3 4 5 6 7 8 9 10 Very

How do you feel about the healthiness and size of this meal?

large/ unhealthy   small/ healthy

☹ ☹ 😐 😊 😄
☐ ☐ ☐ ☐ ☐

Well Done!

Total Lunch Calories:

Snack | Amount | Calories | Time: | Location:

How hungry were you before eating?

Not at all 1 2 3 4 5 6 7 8 9 10 Very

Why did you eat this snack?

Total Snack Calories:

			Weight	
			Time	

Dinner	Amount	Calories	Time:	Location:
Total Dinner Calories:				

How hungry were you before eating?

Not at all 1 2 3 4 5 6 7 8 9 10 **Very**

How do you feel about the healthiness and size of this meal?

large/unhealthy   small/healthy

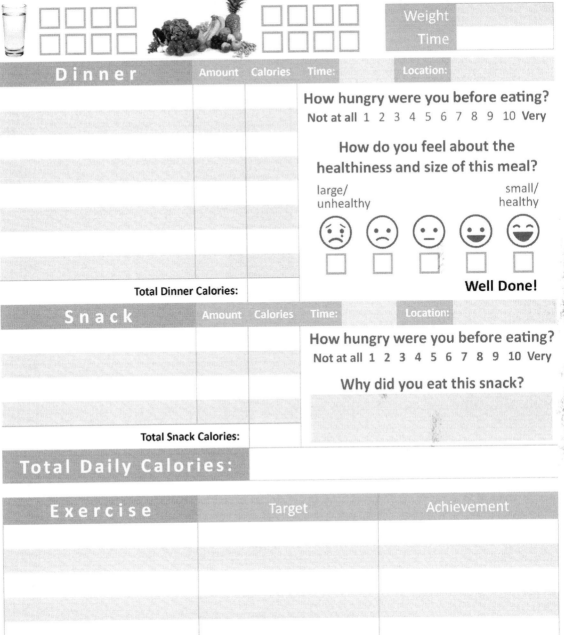

☐ ☐ ☐ ☐ ☐

Well Done!

Snack	Amount	Calories	Time:	Location:
Total Snack Calories:				

How hungry were you before eating?

Not at all 1 2 3 4 5 6 7 8 9 10 **Very**

Why did you eat this snack?

Total Daily Calories:

Exercise	Target	Achievement

Are you happy with how you ate and exercised today?

Food	Exercise
☐ ☐ ☐ ☐ ☐	☐ ☐ ☐ ☐ ☐
Well Done!	**Well Done!**

Your Weekly Progress

Date: _____

	This week's measurements
Weight:	
Chest:	
Waist:	
Hips:	
Thighs:	
Calves:	
Upper arms:	
Cholesterol:	
Blood pressure:	

How do you feel about this week's progress?

☐ ☐ ☐ ☐ ☐

Well Done!

Things you did well this week:

Things you can improve:

Week 10

Monday Week 10

Breakfast	Amount	Calories	Time:	Location:
Total Breakfast Calories:				

How hungry were you before eating?
Not at all 1 2 3 4 5 6 7 8 9 10 Very

How do you feel about the healthiness and size of this meal?

large/unhealthy   small/healthy

☐ ☐ ☐ ☐ ☐

Well Done!

Snack	Amount	Calories	Time:	Location:
Total Snack Calories:				

How hungry were you before eating?
Not at all 1 2 3 4 5 6 7 8 9 10 Very

Why did you eat this snack?

Lunch	Amount	Calories	Time:	Location:
Total Lunch Calories:				

How hungry were you before eating?
Not at all 1 2 3 4 5 6 7 8 9 10 Very

How do you feel about the healthiness and size of this meal?

large/unhealthy   small/healthy

☐ ☐ ☐ ☐ ☐

Well Done!

Snack	Amount	Calories	Time:	Location:
Total Snack Calories:				

How hungry were you before eating?
Not at all 1 2 3 4 5 6 7 8 9 10 Very

Why did you eat this snack?

Weight

Time

Dinner

	Amount	Calories	Time:	Location:

How hungry were you before eating?

Not at all 1 2 3 4 5 6 7 8 9 10 Very

How do you feel about the healthiness and size of this meal?

large/unhealthy   small/healthy

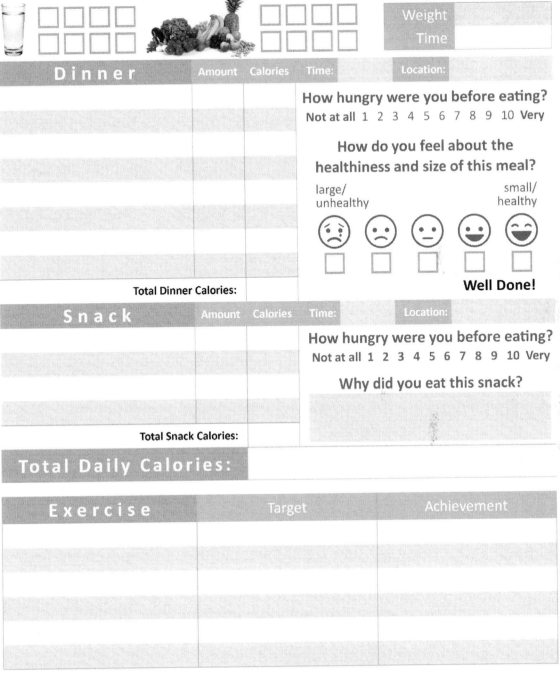

Well Done!

Total Dinner Calories:

Snack

	Amount	Calories	Time:	Location:

How hungry were you before eating?

Not at all 1 2 3 4 5 6 7 8 9 10 Very

Why did you eat this snack?

Total Snack Calories:

Total Daily Calories:

Exercise	Target	Achievement

Are you happy with how you ate and exercised today?

Food	Exercise

Well Done!   **Well Done!**

Tuesday Week 10

Breakfast

	Amount	Calories	Time:	Location:

Total Breakfast Calories:

How hungry were you before eating?

Not at all 1 2 3 4 5 6 7 8 9 10 Very

How do you feel about the healthiness and size of this meal?

large/ unhealthy   small/ healthy

☐ ☐ ☐ ☐ ☐

Well Done!

Snack

	Amount	Calories	Time:	Location:

Total Snack Calories:

How hungry were you before eating?

Not at all 1 2 3 4 5 6 7 8 9 10 Very

Why did you eat this snack?

Lunch

	Amount	Calories	Time:	Location:

Total Lunch Calories:

How hungry were you before eating?

Not at all 1 2 3 4 5 6 7 8 9 10 Very

How do you feel about the healthiness and size of this meal?

large/ unhealthy   small/ healthy

☐ ☐ ☐ ☐ ☐

Well Done!

Snack

	Amount	Calories	Time:	Location:

Total Snack Calories:

How hungry were you before eating?

Not at all 1 2 3 4 5 6 7 8 9 10 Very

Why did you eat this snack?

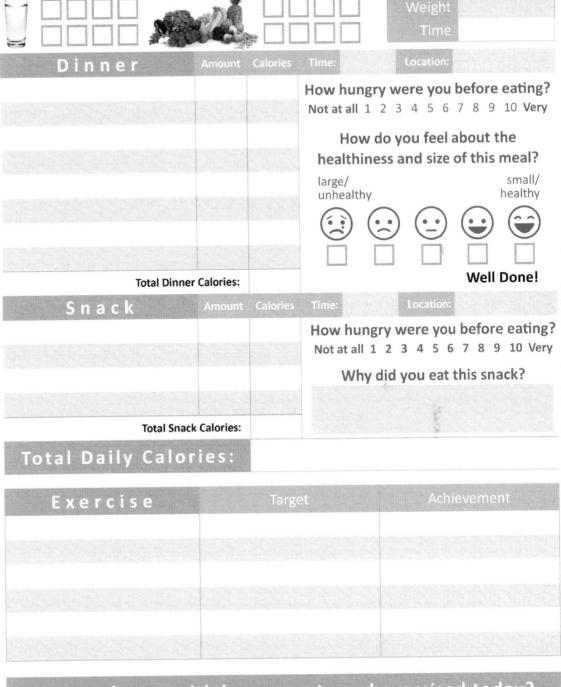

	Weight
	Time

Dinner

	Amount	Calories	Time:	Location:
Total Dinner Calories:				

How hungry were you before eating?

Not at all 1 2 3 4 5 6 7 8 9 10 Very

How do you feel about the healthiness and size of this meal?

large/unhealthy · · · · · · · · · · · · · · · · · · small/healthy

☐ ☐ ☐ ☐ ☐

Well Done!

Snack

	Amount	Calories	Time:	Location:
Total Snack Calories:				

How hungry were you before eating?

Not at all 1 2 3 4 5 6 7 8 9 10 Very

Why did you eat this snack?

Total Daily Calories:

Exercise	Target	Achievement

Are you happy with how you ate and exercised today?

Food	Exercise
☐ ☐ ☐ ☐ ☐	☐ ☐ ☐ ☐ ☐
Well Done!	**Well Done!**

Wednesday Week 10

Breakfast | Amount | Calories | Time: | Location:

	Amount	Calories	Time:	Location:

How hungry were you before eating?
Not at all 1 2 3 4 5 6 7 8 9 10 Very

How do you feel about the healthiness and size of this meal?

large/ unhealthy   small/ healthy

☐ ☐ ☐ ☐ ☐

Well Done!

Total Breakfast Calories:

Snack | Amount | Calories | Time: | Location:

	Amount	Calories	Time:	Location:

How hungry were you before eating?
Not at all 1 2 3 4 5 6 7 8 9 10 Very

Why did you eat this snack?

Total Snack Calories:

Lunch | Amount | Calories | Time: | Location:

	Amount	Calories	Time:	Location:

How hungry were you before eating?
Not at all 1 2 3 4 5 6 7 8 9 10 Very

How do you feel about the healthiness and size of this meal?

large/ unhealthy   small/ healthy

☐ ☐ ☐ ☐ ☐

Well Done!

Total Lunch Calories:

Snack | Amount | Calories | Time: | Location:

	Amount	Calories	Time:	Location:

How hungry were you before eating?
Not at all 1 2 3 4 5 6 7 8 9 10 Very

Why did you eat this snack?

Total Snack Calories:

		Weight	
		Time	

Dinner

	Amount	Calories	Time:	Location:
Total Dinner Calories:				

How hungry were you before eating?

Not at all 1 2 3 4 5 6 7 8 9 10 Very

How do you feel about the healthiness and size of this meal?

large/unhealthy   small/healthy

☹ 😕 😐 😃 😄

☐ ☐ ☐ ☐ ☐

Well Done!

Snack

	Amount	Calories	Time:	Location:
Total Snack Calories:				

How hungry were you before eating?

Not at all 1 2 3 4 5 6 7 8 9 10 Very

Why did you eat this snack?

Total Daily Calories:

Exercise	Target	Achievement

Are you happy with how you ate and exercised today?

Food	Exercise
☹ 😕 😐 😃 😄	☹ 😕 😐 😃 😄
☐ ☐ ☐ ☐ ☐	☐ ☐ ☐ ☐ ☐
Well Done!	**Well Done!**

Thursday Week 10

Breakfast	Amount	Calories	Time:	Location:
Total Breakfast Calories:				

How hungry were you before eating?
Not at all 1 2 3 4 5 6 7 8 9 10 Very

How do you feel about the healthiness and size of this meal?

large/
unhealthy

small/
healthy

☹ ☹ 😐 🙂 😄
☐ ☐ ☐ ☐ ☐

Well Done!

Snack	Amount	Calories	Time:	Location:
Total Snack Calories:				

How hungry were you before eating?
Not at all 1 2 3 4 5 6 7 8 9 10 Very

Why did you eat this snack?

Lunch	Amount	Calories	Time:	Location:
Total Lunch Calories:				

How hungry were you before eating?
Not at all 1 2 3 4 5 6 7 8 9 10 Very

How do you feel about the healthiness and size of this meal?

large/
unhealthy

small/
healthy

☹ ☹ 😐 🙂 😄
☐ ☐ ☐ ☐ ☐

Well Done!

Snack	Amount	Calories	Time:	Location:
Total Snack Calories:				

How hungry were you before eating?
Not at all 1 2 3 4 5 6 7 8 9 10 Very

Why did you eat this snack?

					Weight	
					Time	

Dinner

	Amount	Calories	Time:	Location:

Total Dinner Calories:

How hungry were you before eating?

Not at all 1 2 3 4 5 6 7 8 9 10 Very

How do you feel about the healthiness and size of this meal?

large/unhealthy   small/healthy

☹ ☹ 😐 🙂 😄
☐ ☐ ☐ ☐ ☐

Well Done!

Snack

	Amount	Calories	Time:	Location:

Total Snack Calories:

How hungry were you before eating?

Not at all 1 2 3 4 5 6 7 8 9 10 Very

Why did you eat this snack?

Total Daily Calories:

Exercise	Target	Achievement

Are you happy with how you ate and exercised today?

Food	Exercise
☹ ☹ 😐 🙂 😄	☹ ☹ 😐 🙂 😄
☐ ☐ ☐ ☐ ☐	☐ ☐ ☐ ☐ ☐
Well Done!	**Well Done!**

Friday Week 10

Breakfast

	Amount	Calories	Time:	Location:

How hungry were you before eating?

Not at all 1 2 3 4 5 6 7 8 9 10 Very

How do you feel about the healthiness and size of this meal?

large/unhealthy   small/healthy

☹ ☹ 😐 🙂 😄

☐ ☐ ☐ ☐ ☐

Well Done!

Total Breakfast Calories:

Snack

	Amount	Calories	Time:	Location:

How hungry were you before eating?

Not at all 1 2 3 4 5 6 7 8 9 10 Very

Why did you eat this snack?

Total Snack Calories:

Lunch

	Amount	Calories	Time:	Location:

How hungry were you before eating?

Not at all 1 2 3 4 5 6 7 8 9 10 Very

How do you feel about the healthiness and size of this meal?

large/unhealthy   small/healthy

☹ ☹ 😐 🙂 😄

☐ ☐ ☐ ☐ ☐

Well Done!

Total Lunch Calories:

Snack

	Amount	Calories	Time:	Location:

How hungry were you before eating?

Not at all 1 2 3 4 5 6 7 8 9 10 Very

Why did you eat this snack?

Total Snack Calories:

Weight	
Time	

D i n n e r

	Amount	Calories	Time:	Location:

How hungry were you before eating?

Not at all 1 2 3 4 5 6 7 8 9 10 **Very**

How do you feel about the healthiness and size of this meal?

large/
unhealthy   small/
healthy

☹ 🙁 😐 🙂 😄

☐ ☐ ☐ ☐ ☐

Well Done!

Total Dinner Calories:

S n a c k

	Amount	Calories	Time:	Location:

How hungry were you before eating?

Not at all 1 2 3 4 5 6 7 8 9 10 **Very**

Why did you eat this snack?

Total Snack Calories:

Total Daily Calories:

Exercise	Target	Achievement

Are you happy with how you ate and exercised today?

Food	Exercise
☹ 🙁 😐 🙂 😄	☹ 🙁 😐 🙂 😄
☐ ☐ ☐ ☐ ☐	☐ ☐ ☐ ☐ ☐
Well Done!	**Well Done!**

Saturday Week 10

Breakfast

Amount	Calories	Time:	Location:

How hungry were you before eating?

Not at all 1 2 3 4 5 6 7 8 9 10 Very

How do you feel about the healthiness and size of this meal?

large/unhealthy   small/healthy

☹ ☹ 😐 🙂 😄

☐ ☐ ☐ ☐ ☐

Total Breakfast Calories: **Well Done!**

Snack

Amount	Calories	Time:	Location:

How hungry were you before eating?

Not at all 1 2 3 4 5 6 7 8 9 10 Very

Why did you eat this snack?

Total Snack Calories:

Lunch

Amount	Calories	Time:	Location:

How hungry were you before eating?

Not at all 1 2 3 4 5 6 7 8 9 10 Very

How do you feel about the healthiness and size of this meal?

large/unhealthy   small/healthy

☹ ☹ 😐 🙂 😄

☐ ☐ ☐ ☐ ☐

Total Lunch Calories: **Well Done!**

Snack

Amount	Calories	Time:	Location:

How hungry were you before eating?

Not at all 1 2 3 4 5 6 7 8 9 10 Very

Why did you eat this snack?

Total Snack Calories:

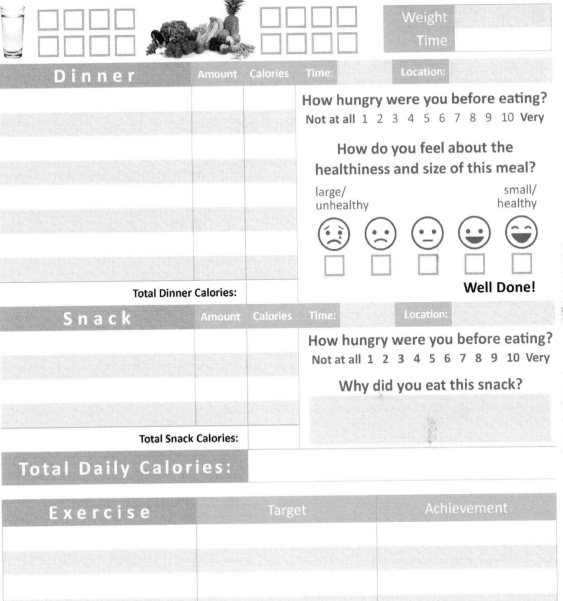

		Weight	
		Time	

Dinner

	Amount	Calories	Time:	Location:

How hungry were you before eating?

Not at all 1 2 3 4 5 6 7 8 9 10 Very

How do you feel about the healthiness and size of this meal?

large/unhealthy   small/healthy

☐ ☐ ☐ ☐ ☐

Well Done!

Total Dinner Calories:

Snack

	Amount	Calories	Time:	Location:

How hungry were you before eating?

Not at all 1 2 3 4 5 6 7 8 9 10 Very

Why did you eat this snack?

Total Snack Calories:

Total Daily Calories:

Exercise	Target	Achievement

Are you happy with how you ate and exercised today?

Food	Exercise
☐ ☐ ☐ ☐ ☐	☐ ☐ ☐ ☐ ☐
Well Done!	**Well Done!**

Sunday Week 10

Breakfast | Amount | Calories | Time: | Location:

How hungry were you before eating?

Not at all 1 2 3 4 5 6 7 8 9 10 Very

How do you feel about the healthiness and size of this meal?

large/
unhealthy   small/
healthy

☹ ☹ 😐 🙂 😄

☐ ☐ ☐ ☐ ☐

Well Done!

Total Breakfast Calories:

Snack | Amount | Calories | Time: | Location:

How hungry were you before eating?

Not at all 1 2 3 4 5 6 7 8 9 10 Very

Why did you eat this snack?

Total Snack Calories:

Lunch | Amount | Calories | Time: | Location:

How hungry were you before eating?

Not at all 1 2 3 4 5 6 7 8 9 10 Very

How do you feel about the healthiness and size of this meal?

large/
unhealthy   small/
healthy

☹ ☹ 😐 🙂 😄

☐ ☐ ☐ ☐ ☐

Well Done!

Total Lunch Calories:

Snack | Amount | Calories | Time: | Location:

How hungry were you before eating?

Not at all 1 2 3 4 5 6 7 8 9 10 Very

Why did you eat this snack?

Total Snack Calories:

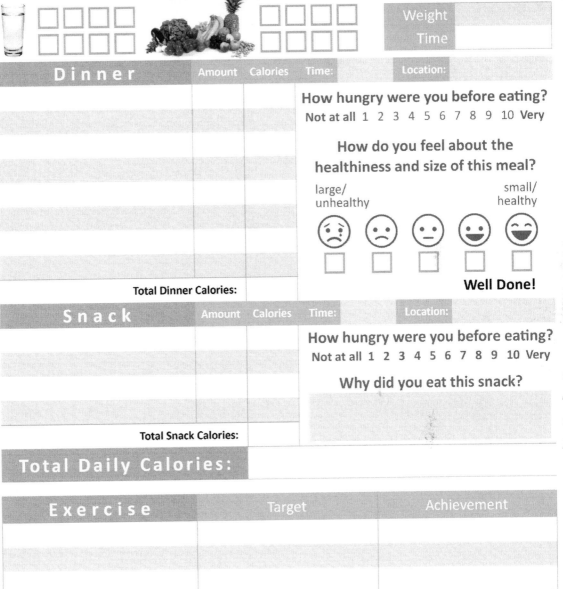

| Weight | |
| Time | |

D i n n e r

	Amount	Calories	Time:	Location:

How hungry were you before eating?

Not at all 1 2 3 4 5 6 7 8 9 10 **Very**

How do you feel about the healthiness and size of this meal?

large/unhealthy   small/healthy

Well Done!

Total Dinner Calories:

S n a c k

	Amount	Calories	Time:	Location:

How hungry were you before eating?

Not at all 1 2 3 4 5 6 7 8 9 10 **Very**

Why did you eat this snack?

Total Snack Calories:

Total Daily Calories:

Exercise	Target	Achievement

Are you happy with how you ate and exercised today?

Food	Exercise

Well Done!   **Well Done!**

Your Weekly Progress

Date: _____

	This week's measurements
Weight:	
Chest:	
Waist:	
Hips:	
Thighs:	
Calves:	
Upper arms:	
Cholesterol:	
Blood pressure:	

How do you feel about this week's progress?

☹ ☐ 🙁 ☐ 😐 ☐ 🙂 ☐ 😄 ☐

Well Done!

Things you did well this week:

Things you can improve:

Printed in Great Britain
by Amazon